AF595101

Bioassays and Biosensors for Rapid Detection and Analysis

Bioassays and Biosensors for Rapid Detection and Analysis

Editor

Zhen Zhang

Basel • Beijing • Wuhan • Barcelona • Belgrade • Novi Sad • Cluj • Manchester

Editor
Zhen Zhang
Jiangsu University
Zhenjiang, China

Editorial Office
MDPI
St. Alban-Anlage 66
4052 Basel, Switzerland

This is a reprint of articles from the Special Issue published online in the open access journal *Biosensors* (ISSN 2079-6374) (available at: https://www.mdpi.com/journal/biosensors/special_issues/bioassay).

For citation purposes, cite each article independently as indicated on the article page online and as indicated below:

Lastname, A.A.; Lastname, B.B. Article Title. *Journal Name* **Year**, *Volume Number*, Page Range.

ISBN 978-3-0365-9082-0 (Hbk)
ISBN 978-3-0365-9083-7 (PDF)
doi.org/10.3390/books978-3-0365-9083-7

Contents

About the Editor

Zhen Zhang

Zhen Zhang, he graduated from China Agricultural University with a bachelor's degree or a master's degree respectively in 2001 and 2006. He entered the ecological environment research center of the Chinese Academy of Sciences to study for a doctor's degree in 2007, and joined the school of environmental and safety engineering of Jiangsu University with a doctor's degree in 2010. He was a postdoctoral fellow at Changwon National University in Korea from 2012 to 2013, and a visiting scholar at North Carolina State University in the United States from 2016 to 2017. The research direction involves the establishment of rapid analysis methods for pollutants, equipment development and environmental behavior and risk assessment; Exploration of new bioanalytical methods based on nucleic acid functional materials and their applications in environment, food and health.

 biosensors 

Editorial

Recent Progress in Biosensors Based on Biorecognition Molecules

Zhen Zhang

School of the Environment and Safety Engineering, Jiangsu University, Zhenjiang 212013, China; zhangzhen@ujs.edu.cn

Biosensors are considered a popular technology to rapidly detect targets, and are generally composed of biorecognition molecules that specifically capture analytes and signal elements. In terms of methodology, measurements are carried out in a limited time without sophisticated sample pre-treatments, implying obvious advantages over traditional instrument methods for rapid analysis [1].

Considering that biosensors are widely applied in environmental pollutant monitoring, detecting hazardous substances in foods and disease diagnosis [2–4], their advances and future trends should be highlighted, so we have organized this Special Issue entitled "Bioassays and Biosensors for Rapid Detection and Analysis". Among the 10 papers published, 6 are articles, 3 are communications and 1 is a review.

Huang et al. believe that electrogenerated chemiluminescence (ECL) is a powerful tool for sensitive and accurate detection of biological analytes and summarize the recent advances in this field. Moreover, some future trends and challenges of ECL biosensors are also discussed. They believe that after integration with low-cost photodetectors, ECL biosensors will have great potential in commercial applications [5]. Meanwhile, several works focus on contaminant determination in food and environments, like sulfadiazine (SDZ) [6], deoxynivalenol (DON) [7] and dibutyl phthalate (DBP) [8]. With regard to the detection systems, antibodies or aptamers were used to recognize the targets, which contributed to good selectivity. In addition, some novel strategies for signal amplification were employed to improve their analytical performance [8]. At the same time, sensitive biosensors were fabricated for the rapid detection of the biomarkers related to some diseases, which could be achieved easily [9,10].

Most studies focus on the sensitivity improvement of biosensors through introducing new functional materials or simplifying the measurement procedure, which are definitely important. However, more attention should be paid to real applications, increasing tolerances against various matrices that potentially influence the performance of biosensors. Interestingly, in this Special Issue, two papers are related to commercial products using bioassays or biosensors [11,12]. In short, biosensors will be widely used in various fields after some problems have been overcome, including achieving satisfactory sensitivity and accuracy and maintaining enzyme activity in varied harsh environments.

Funding: This research received no external funding.

Conflicts of Interest: The authors declare no conflict of interest.

Citation: Zhang, Z. Recent Progress in Biosensors Based on Biorecognition Molecules. *Biosensors* **2023**, *13*, 842. https://doi.org/10.3390/bios13090842

Received: 21 August 2023
Accepted: 23 August 2023
Published: 24 August 2023

References

1. Zhang, Z.; Zeng, K.; Liu, J. Immunochemical detection of emerging organic contaminants in environmental waters. *TrAC Trends Anal. Chem.* **2017**, *87*, 49–57. [CrossRef]
2. Pilolli, R.; Monaci, L.; Visconti, A. Advances in biosensor development based on integrating nanotechnology and applied to food-allergen management. *TrAC Trends Anal. Chem.* **2013**, *47*, 12–26. [CrossRef]
3. Iwanaga, M.; Tangkawsakul, W. Two-Way Detection of COVID-19 Spike Protein and Antibody Using All-Dielectric Metasurface Fluorescence Sensors. *Biosensors* **2022**, *12*, 981. [CrossRef]
4. Niehues, J.; McElroy, C.; Croon, A.; Pietschmann, J.; Frettlöh, M.; Schröper, F. Bacterial Lighthouses— Real-Time Detection of Yersinia enterocolitica by Quorum Sensing. *Biosensors* **2021**, *11*, 517. [PubMed]
5. Huang, Y.; Yao, Y.; Wang, Y.; Chen, L.; Zeng, Y.; Li, L.; Guo, L. Strategies for Enhancing the Sensitivity of Electrochemiluminescence Biosensors. *Biosensors* **2022**, *12*, 750. [CrossRef] [PubMed]
6. Zheng, X.; Yang, L.; Sun, Q.; Zhang, L.; Le, T. Development and Validation of Aptasensor Based on MnO2 for the Detection of Sulfadiazine Residues. *Biosensors* **2023**, *13*, 613. [CrossRef] [PubMed]
7. Zeng, K.; Yang, J.; Su, H.; Yang, S.; Gu, X.; Zhang, Z.; Zhao, H. Enhanced Competitive Immunomagnetic Beads Assay Assisted with PAMAM-Gold Nanoparticles Multi-Enzyme Probes for Detection of Deoxynivalenol. *Biosensors* **2023**, *13*, 536. [CrossRef] [PubMed]
8. Meng, H.; Yao, N.; Zeng, K.; Zhu, N.; Wang, Y.; Zhao, B.; Zhang, Z. A Novel Enzyme-Free Ratiometric Fluorescence Immunoassay Based on Silver Nanoparticles for the Detection of Dibutyl Phthalate from Environmental Waters. *Biosensors* **2022**, *12*, 125. [CrossRef] [PubMed]
9. Alekhmimi, N.K.; Cialla-May, D.; Ramadan, Q.; Eissa, S.; Popp, J.; Al-Kattan, K.; Zourob, M. Biosensing Platform for the Detection of Biomarkers for ALI/ARDS in Bronchoalveolar Lavage Fluid of LPS Mice Model. *Biosensors* **2023**, *13*, 676. [CrossRef] [PubMed]
10. Khan, A.; Di, K.; Khan, H.; He, N.; Li, Z. Rapid Capturing and Chemiluminescent Sensing of Programmed Death Ligand-1 Expressing Extracellular Vesicles. *Biosensors* **2022**, *12*, 281. [CrossRef] [PubMed]
11. Dobrynin, D.; Polischuk, I.; Pokroy, B. A Comparison Study of the Detection Limit of Omicron SARS-CoV-2 Nucleocapsid by Various Rapid Antigen Tests. *Biosensors* **2022**, *12*, 1083. [CrossRef] [PubMed]
12. Johannsen, B.; Baumgartner, D.; Karkossa, L.; Paust, N.; Karpíšek, M.; Bostanci, N.; Zengerle, R.; Mitsakakis, K. ImmunoDisk—A Fully Automated Bead-Based Immunoassay Cartridge with All Reagents Pre-Stored. *Biosensors* **2022**, *12*, 413. [PubMed]

Communication

Development and Validation of Aptasensor Based on MnO_2 for the Detection of Sulfadiazine Residues

Xiaoling Zheng, Lulan Yang, Qi Sun, Lei Zhang and Tao Le *

College of Life Sciences, Chongqing Normal University, Chongqing 401331, China;
2021110513064@stu.cqnu.edu.cn (X.Z.); 2020110513031@stu.cqnu.edu.cn (L.Y.);
sunqi2017@cqnu.edu.cn (Q.S.); 20132133@cqnu.edu.cn (L.Z.)
* Correspondence: letao@cqnu.edu.cn

Abstract: The monitoring of sulfadiazine (SDZ) is of great significance for food safety, environmental protection, and human health. In this study, a fluorescent aptasensor based on MnO_2 and FAM-labeled SDZ aptamer (FAM-SDZ30-1) was developed for the sensitive and selective detection of SDZ in food and environmental samples. MnO_2 nanosheets adsorbed rapidly to the aptamer through its electrostatic interaction with the base, providing the basis for an ultrasensitive SDZ detection. Molecular dynamics was used to explain the combination of SMZ1S and SMZ. This fluorescent aptasensor exhibited high sensitivity and selectivity with a limit of detection of 3.25 ng/mL and a linear range of 5–40 ng/mL. The recoveries ranged from 87.19% to 109.26% and the coefficients of variation ranged from 3.13% to 13.14%. In addition, the results of the aptasensor showed an excellent correlation with high-performance liquid chromatography (HPLC). Therefore, this aptasensor based on MnO_2 is a potentially useful methodology for highly sensitive and selective detection of SDZ in foods and environments.

Keywords: aptamer; MnO_2; aptasensor; sulfadiazine

Citation: Zheng, X.; Yang, L.; Sun, Q.; Zhang, L.; Le, T. Development and Validation of Aptasensor Based on MnO_2 for the Detection of Sulfadiazine Residues. *Biosensors* **2023**, *13*, 613. https://doi.org/10.3390/bios13060613

Received: 27 April 2023
Revised: 27 May 2023
Accepted: 1 June 2023
Published: 3 June 2023

1. Introduction

Sulfadiazine (SDZ) is an antibiotic of the sulfonamide family, which is widely applied to prevent and treat bacterial infections in livestock [1,2]. However, SDZ can lead to residues in animal-derived foods when overused. These residues can not only accumulate in humans through the food chain and cause serious health problems, but can also pollute soil or water by the pathway of animal excretion, leading to ecotoxicological contamination [3,4]. After discovering its harmful effects, the maximum residue level of SDZ has been set at 100 μg/kg in the EU, the USA, and China. Therefore, monitoring the residues of SDZ in food and the environment is of great significance [5,6]. In previous studies, SDZ was detected by HPLC, capillary electrophoresis, enzyme-linked immunosorbent assay, and time-resolved immunoassay [7–11]. Although these approaches are highly sensitive, the necessity of specialized operators, high costs, weak stability of antibodies, and the potential of cross-reactivity limit their application. Therefore, a simple, effective, and economical method is required to measure SDZ in food and environmental risk monitoring.

Aptamer development as a new type of recognition probe provides new ideas for the development of its detection. Aptamers are single-stranded DNA (ssDNA) or RNA that specifically bind to target molecules. Aptamers have superior stability, excellent affinity, strong specificity, and easy in vitro screening compared to antibodies [12–14]. With these advantages, various aptamer-based sensors have been widely developed for environmental monitoring and food safety inspection (e.g., fluorescent, electrochemical, surface-enhanced Raman scattering, and photoelectrochemical sensing signals) [15–17]. In comparison, fluorescent aptasensors have been widely used in the field of detection for their ability of sensitivity, low cost, and selectivity, which make them particularly useful in monitoring food and environmental samples [18,19], and fluorescent aptamer sensors based on nanomaterials (quantum dots, graphene oxide, nanosilicon, etc.) have attracted considerable attraction [20–22]. In particular, manganese dioxide (MnO_2) has attracted much attention for its low cost, strong molar extinction coefficient, environmental tolerance, and non-toxicity, which has been widely used in biomedical and biosensing fields [22,23]. As a constantly developing two-dimensional nanomaterial, MnO_2 nanosheets are easy to prepare on a large scale and possess broad absorption peaks that can overlap with the emission spectra of many fluorescent materials, making them considered to be quencher candidates. In addition, because MnO_2 has a large specific surface area and good biocompatibility, it can greatly improve the detection of targets. Xu et al. reported that MnO_2 nanosheets adsorb ssDNA through van der Waals forces interacting with bases, thus exhibiting strong absorption and efficiency of fluorescence burst [24]. Consequently, Qin et al. established an upconversion MnO_2-based biosensor for the detection of carbendazim pesticides in food [25]. Li et al. constructed an open fluorescence sensor based on carbon dot-labeled oligodeoxyribonucleotide and MnO_2 nanosheets for the detection of mercury(II) [26]. However, MnO_2-based fluorescent aptasensors for the detection of SDZ residues have not been reported so far.

In this research, we constructed an aptasensor for selective detection of SDZ using FAM-labeled aptamer (FAM-SDZ30-1) and MnO_2 nanosheets as donor-acceptor pair for the first time. Molecular dynamics simulations were performed to study the reason for the high specificity for SDZ by SDZ30-1. The performance of the established aptasensor was tested in food and environmental samples by adding different concentrations of SDZ to five samples, including soil, lake water, river water, egg, and beef. Finally, we compared the results of this method with HPLC and explored their correlation, demonstrating that the aptasensor can detect SDZ in a variety of samples.

2. Materials and Methods

2.1. Chemicals and Materials

The SDZ 30-1 (5′-AACCCAATGGGAT-3′, K_d = 65.72 nM) was selected in our laboratory, which was purified using HPLC and synthesized by Sangon Biotech, Shanghai, China [24]. All antibiotics, e.g., SDZ, tetracycline (TC), furaltadone (FTD), and norfloxacin (OFL) were bought from Sigma-Aldrich, St. Louis, MO, USA. $KMnO_4$, hexadecyl trimethyl ammonium bromide (CTAB), buffer, and other reagents were purchased from Aladdin Co., Ltd., Ontario, CA, USA. The accuracy of the method was verified by HPLC on LC-20A (Shimadzu, Kyoto, Japan). Fluorescence intensity measurements were performed at the excitation wavelength of 492 nm and emission wavelength of 518 nm using Varioskan LUX.

2.2. Synthesis of MnO_2

The MnO_2 nanosheets were prepared as previously described [23]. Specifically, 0.5 g $KMnO_4$ was dissolved in 450 mL of water and stirred for 30 min, then 1.5 g of CATB was added to form a stable emulsion. Next, 50 mL of 0.1 M MES (pH 6.0) was added to the above mixture and stirred for 12 h until a black color was formed. The product was centrifuged at 8000 rpm for 10 min and then washed thrice alternately with water and alcohol. Finally, the collected product was dried under a vacuum at 60 °C for 12 h. MnO_2 was characterized by transmission electron microscope images (TEM), the UV-Vis, EDS elemental mapping and Fourier transform infrared spectrometer (FTIR).

2.3. Molecular Dynamics Simulation

The interactions between aptamers and different targets were compared through molecular dynamics simulations [27,28]. Firstly, the structures and parameters of the aptamer and target molecules were obtained from SwissParam. Subsequently, implicit solvation simulations were performed for the aptamer to target under constant temperature, volume, pressure, etc., while the obtained structures were subjected to an explicit solvent model of molecular dynamics simulations. Finally, the stability was measured using the root mean squared deviation (RMSD) produced by the molecular dynamics simulations trajectory. When the RMSD curve showed a slope upward tendency, the conformation of the system might undergo significant movement, and the smooth oscillation of the RMSD curve around a certain height indicated that the system had reached equilibrium. Other than that, a lower RMSD value means that the structural deviation of the binomial has small and higher stability.

2.4. Optimization of Aptasensor Conditions

We optimized the detection conditions to obtain high sensitivity and selectivity with the established aptasensor. First, to determine the optimal concentration of FAM-SDZ30-1 with MnO_2, different concentrations of MnO_2 (0, 25, 50, 75, 100, 125, 150, 175, 200 μg/mL) were incubated with 50 nM aptamer. After centrifugation, the fluorescence intensity of the supernatant was measured, and the optimal concentration of MnO_2 was determined. Subsequently, at the optimal MnO_2 concentration, different concentrations of aptamer (50, 100, 200, 400, 600, 800 nM) were incubated with 100 ng/mL SDZ, followed by the addition of MnO_2 to the mixture, incubation, and centrifugation to determine the optimal aptamer concentration based on fluorescence intensity. To identify the optimal reaction time of this sensor, MnO_2 was added to the FAM-SDZ30-1 and incubated for 0, 2, 4, 6, 8, 10, 12, and 14 min, and the optimal time of the aptamer with MnO_2 was determined by the value of fluorescence. To find the effects of the reaction time of SDZ with the FAM-aptamer on the fluorescence intensity, the aptamer was then incubated with SDZ for 10, 20, 30, 40, 50, and 60 min. The fluorescence intensity of the supernatant was measured, and the best incubation time of the aptamer with the target was determined. Finally, to explore the effects of the binding buffer on the aptasensor, 400 nM of FAM-SDZ30-1 was incubated with 100 ng/mL of SDZ with four different buffers (HEPES, TE, PBS, Tris-HCl), followed by the addition of 150 μg/mL of MnO_2, incubated and centrifuged, resulting in the determination of the optimized buffer according to the fluorescence in the supernatant. Similarly, the aptasensor was kept at different pH values (6.4, 6.9, 7.4, 7.9, 8.4) and operated as described above to determine the optimal pH based on the fluorescence intensity.

2.5. Standard Curves of the Aptasensor

Briefly, 40 μL of FAM-labeled aptamer (2 μM) was mixed with SDZ at various concentrations (5, 10, 20, 40, 80, 100, 200, and 400 ng/mL) after gentle shaking for 50 min at 25 °C in the dark. After that, 30 μL MnO_2 (1 mg/mL) was added, and the mixture was incubated for 10 min and centrifuged at 10,000 rpm for 5 min. The fluorescence intensity of the supernatant was measured and a standard curve was established with different SDZ concentrations. The limit of detection (LOD) was acquired by calculating the 3SD/slope (the slope of the linear curve), and the SD was the standard deviation from the fluorescence values of a blank sample [29].

2.6. Determination of Selectivity

The selectivity was evaluated by adding 10 μL (100 ng/mL) of different structurally related compounds, e.g., sulfaquinoxaline (SQX), sulfamethazine (SMZ), tetracycline (TC), chlortetracycline (CTC), nitrofurazone (NFZ), furaltadone (FTD), norfloxacin (NOR) dissolved in 400 nM FAM-SDZ30-1 for 50 min to individual aptasensor against SDZ. After centrifugation, the fluorescence intensity was measured in the supernatant. After taking measurements for different antibiotics with relative fluorescence intensity F and the blank sample with fluorescence intensity F_0, the ΔF ($\Delta F = F - F_0$) was calculated [30].

2.7. Sample Preparation

All samples were prepared as previously described with slight modifications [31]. To prepare soil samples, 2 g of soil was diluted 10 times with PBS (pH 7.4), centrifuged at 13,000 rpm for 20 min, and filtered through a 0.22 μm filter membrane. Lake and river water samples were prepared and diluted 10 times as described above; 2 g of eggs were completely mixed with 4 mL of ethyl acetate and vortexed for 10 min, the supernatant after centrifugation was evaporated under a stream of nitrogen at 40 °C to remove the ethyl acetate from the mixture and finally dissolved in the buffer. Next, 5 g beef was homogenized in a homogenizer, then 25 mL of acetonitrile was added, and the mixture was vortexed and shaken for 15 min, ultrasonicated for 10 min, and then centrifuged at 12,000 rpm for 15 min. The supernatant was moved to 30 mL of acetonitrile-saturated hexane and constantly stirred for 10 min to clear the fat. The organic solvent was removed at 80 °C in a water bath, and the residue was dissolved in 5 mL of binding buffer, diluted 10 times, and filtered through a membrane.

2.8. Assay Validation

Five different samples were used to validate the properties of the aptasensor. First, HPLC was used to confirm that all samples did not have SDZ, and then SDZ was added to the processed sample buffer. The standard curves of the sample matrices were constructed according to the aptasensor method described above. The precision and accuracy of the fluorescent aptasensor were verified by analyzing the samples mentioned above with different SDZ concentrations (10, 20, 30 μg/mL), testing five times for each concentration. To test the credibility of the aptasensor in food and environmental samples, the results of the aptasensor and HPLC were compared using the same samples. Following the established procedure, the HPLC analysis was performed on all spiked samples containing different SDZ concentrations. The linear regression was used to calculate the correlation between the results of aptasensor and HPLC. Recoveries were calculated as (measured concentration/known concentration) × 100%, and the coefficient of variation was determined [32]. Each spiked sample was analyzed five times.

3. Results and Discussion

3.1. Principle of the Aptasensor

The FAM-SDZ30-1 fluorescence was quenched by MnO_2 through π-π stacking, electrostatic adsorption, and electron-induced transfer in the absence of SDZ30-1 (Figure 1). In contrast, in the presence of SDZ, FAM-SDZ30-1 selectively combines with SDZ instead of being adsorbed by MnO_2, preventing FAM-SDZ30-1 fluorescence quenching.

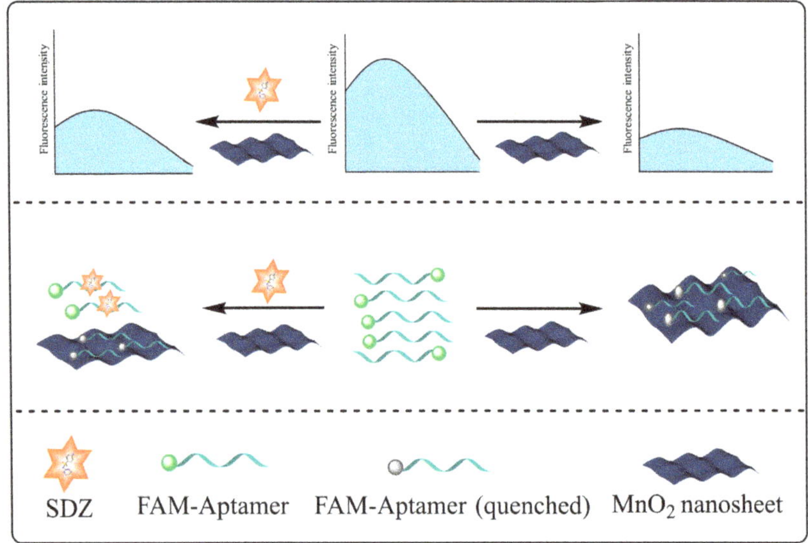

Figure 1. Schematic diagram of the aptasensor based on MnO_2 for the detection of SDZ.

Specifically, the absorption spectrum of MnO_2 and the fluorescence emission spectrum of FAM overlap as MnO_2 is near FAM-SDZ30-1, resulting in FRET (Figure 2A). With the increase of the concentrations of MnO_2, the fluorescence intensity of the aptamer continuously decreased. When the concentration of MnO_2 was extremely high, the fluorescence intensity of the aptamer tended to 0, indicating that the fluorescence of the aptamer was completely quenched (Figure 2B). The fluorescence intensity of FAM in different reaction systems is shown in Figure 2C. When only the aptamer was in the system, the fluorescence intensity was the highest, and after the addition of MnO_2, it will FRET with the aptamer to decrease the fluorescence value. When the target was added to the system, the aptamer and the target could bind specifically, recovering the fluorescence. Moreover, the higher concentration of SDZ, the stronger value of fluorescence was detected. Hence, the fluorescent aptasensor for SDZ detection was constructed based on the correlation between SDZ concentration and fluorescence values.

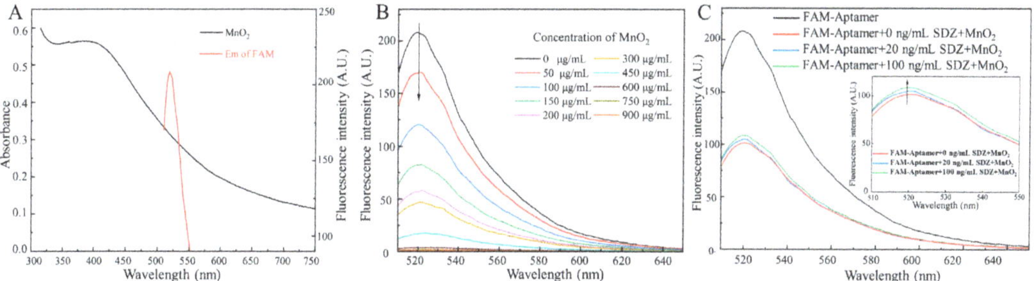

Figure 2. (**A**) UV-Vis absorption spectra of MnO_2 and fluorescence emission spectra of FAM. (**B**) Effects of different MnO_2 concentrations on the fluorescence spectra of FAM. (**C**) Fluorescence intensity of FAM in different reaction systems.

3.2. Characterization of MnO_2

The TEM images showed that the synthesized MnO_2 nanosheets showed a homogeneous structure with a diameter of about 200 nm (Figure 3A). The UV-Vis showed the absorption peak of MnO_2 from 300 to 600 nm (Figure 3B), and the EDS elemental mapping showed that Mn accounted for 68.5% and O 31.5%, demonstrating the successful preparation (Figure 3C). In the MnO_2 curves (Figure 3D), the peaks appearing at 500 to 600 cm^{-1} correspond to Mn-O bending vibration peaks [23]. These results indicated that MnO_2 nanosheets were successfully synthesized.

3.3. Molecular Simulation Analysis

The RMSD curve of the SDZ/SDZ30-1 complex increased slowly from 0 to 15 ns (Figure 4A), indicating that the backbone of the key nucleic acid was moving during this time. Between 15 ns and 100 ns, the complex equilibrated and the RMSD curve fluctuated only around 0.6 Å, with the amplitude remaining within 0.2 Å. In the 0–100 ns range, the RMSD values of the molecular dynamics simulations of the aptamer with other target systems fluctuated strongly, indicating that SDZ30-1 was violently moving with other antibiotics and forming destabilized complexes (Figure 4B–H). Specifically, although SDZ30-1/SMZ reached relative stability at around 20 ns, the amplitude was observed to be large, probably because of their structural similarity. SDZ30-1 with SQX only arrived at relative stability at around 50 ns, yet it continuously increased after 60 ns. The complex formed by SDZ30-1/TC and SDZ30-1/OFL was not balanced until 35 ns the RMSD curve fluctuated around 0.1 Å. Although the complex formed by SDZ30-1 with CTC reached equilibrium at 15 ns, the RMSD curve moved again at about 60 ns. Furthermore, it was evident from the RMSD curves that SDZ30-1/NFZ and SDZ30-1/FTD were not in equivalent equilibrium between 0 and 100 ns, probably because they belong to different antibiotic categories. These results suggested that the aptamer has a high selectivity for SDZ.

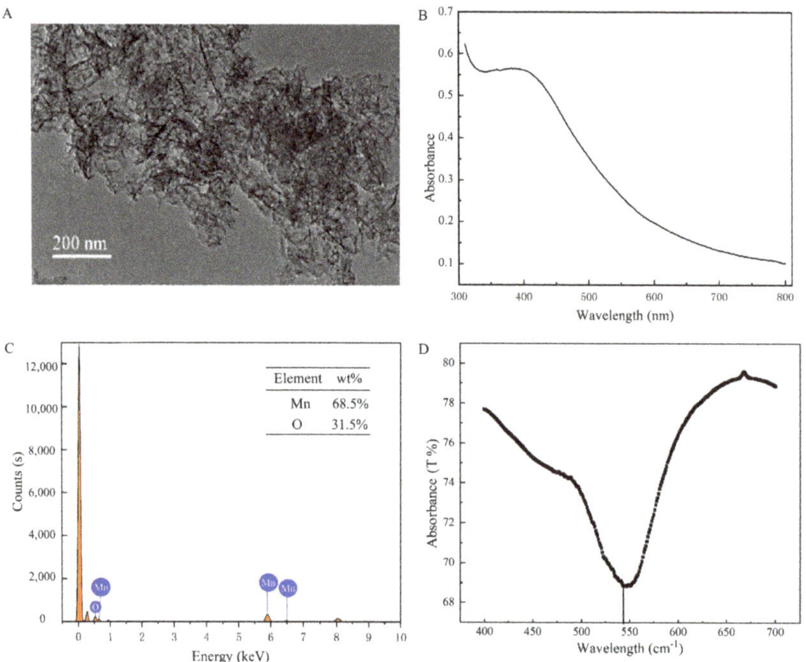

Figure 3. (**A**) TEM images of MnO_2. (**B**) The UV−Vis of MnO_2. (**C**) EDS elemental mapping of MnO_2. (**D**) The FTIR of MnO_2.

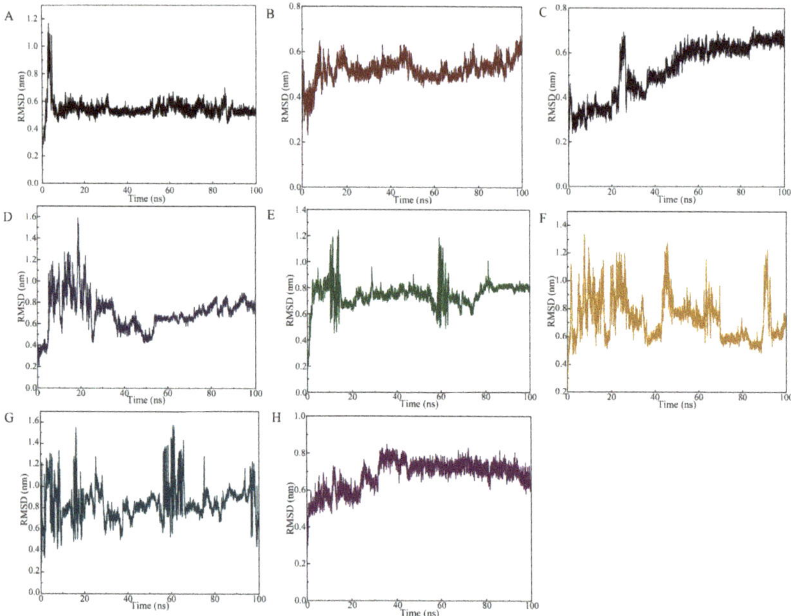

Figure 4. RMSD of the SDZ30-1 aptamer against SDZ and other antibiotics. (**A**) SDZ; (**B**) SMZ; (**C**) SQX; (**D**) tetracycline; (**E**) norfloxacin; (**F**) chlortetracycline; (**G**) nitrofurazone; (**H**) furaltadone.

3.4. Optimization of the Aptasensor

Moreover, we found that as the concentration between FAM-SDZ30-1, MnO_2, and SDZ was optimized, the higher the concentration of MnO_2; the lower the fluorescence intensity of the system, when the concentration reached 150 μg/mL, the fluorescence intensity stabilized, and thus, 150 μg/mL was chosen as the best MnO_2 concentration (Figure 5A). When the concentration of aptamer was 400 nM, the fluorescence value of the system was the highest, and therefore 400 nM was taken as the optimal concentration of the aptamer (Figure 5B). The fluorescence intensity of the aptamer and SDZ reached stability as the time up to 50 min (Figure 5C), thus 50 min was the best incubation time. The fluorescence intensity decreased gradually with the increase of time, and the fluorescence stabilized when the time arrived at 10 min, thereby selected as the best quenching time (Figure 5D). The highest fluorescence intensity of this system was also observed in the PBS buffer (Figure 5E). The pH value affected the fluorescence intensity of the fluorescent aptasensor, and 7.4 was the best pH condition (Figure 5F).

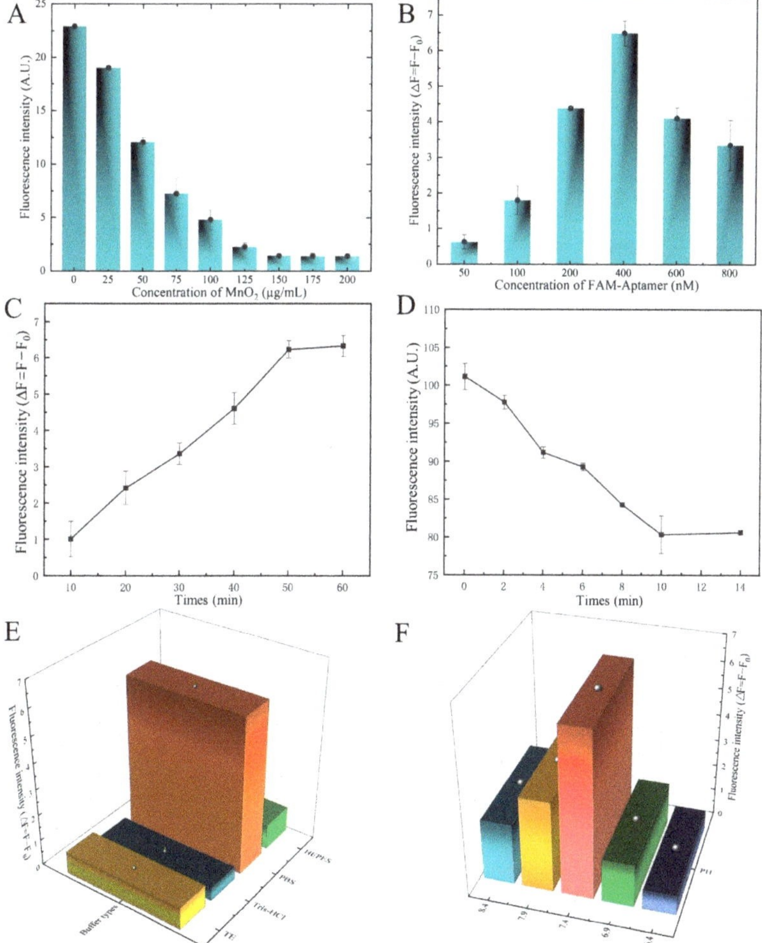

Figure 5. Optimization of the aptasensor system. (**A**) Concentration of MnO_2. (**B**) Concentration of aptamer. (**C**) Incubate time of aptamer with MnO_2. (**D**) Incubate time of aptamer with SDZ. (**E**) Optimization of binding buffer. (**F**) Optimization of pH.

3.5. Properties of Aptasensor

The linear range for SDZ detection varied from 5 to 40 ng/mL with a LOD of 3.25 ng/mL (Figure 6A). Then, we compared this performance to several reported SDZ detection methods. The developed fluorescent aptasensor displayed a similar linear range to traditional methods, such as HPLC, UPLC-MS/MS, and ELISA (Table 1). Compared to electrochemical biosensors, the proposed fluorescence aptasensor avoids complex electrode modifications and is relatively simple to operate. Then, similar fluorescent biosensors based on nanomaterials were compared. The linear range and LOD of the fluorescent biosensor constructed in this study were similar to several previous methods.

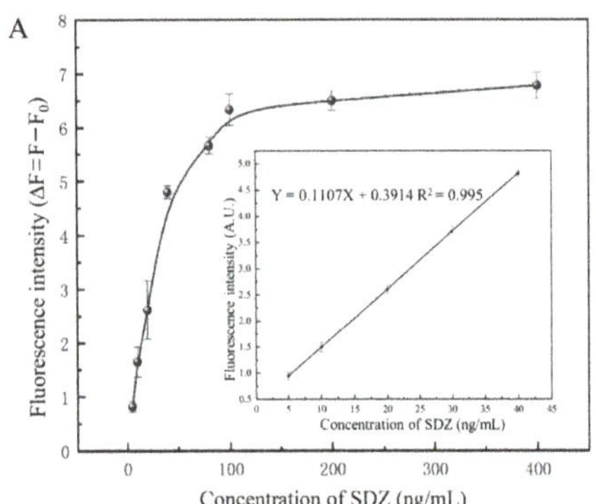

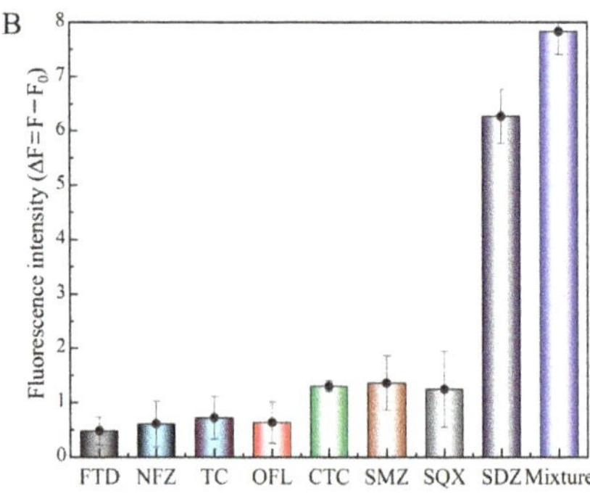

Figure 6. Properties of the fluorescent aptasensor. (**A**) Standard curve of the aptasensor. (**B**) Specificity of SDZ30-1 against SDZ and other antibiotics.

Table 1. Comparison of the method proposed in this study with other methods previously for SDZ.

Method	LOD (ng/mL)	References
HPLC	0.9	Alipanahpour et al. (2021) [8]
Spectrophotometry	19	Errayess et al. (2017) [9]
Chitosan-based ELISA	8.64	Zeng et al. (2021) [10]
Electrochemical	2.25	Kokulnathan et al. (2021) [2]
Aptasensor based on gold nanoparticles	2	Yang et al. (2022) [32]
Aptasensor based on MnO_2	3.25	Present work

Meanwhile, in order to assess the detection selectivity for SDZ of this aptasensor, SDZ, and other antibiotics were detected using the FRET-based aptasensor. The fluorescence intensity of SDZ30-1 bound to all other antibiotics was relatively slow (Figure 6B). Specifically, the fluorescence intensity of SDZ30-1 was similar to SMZ and SQX, probably because they belong to the family of sulfonamide antibiotics and have similar structures. The fluorescence values of SDZ30-1 with FTD, NFZ, TC, and OFL were very low, indicating that the aptasensor has high specificity. Taken together, these results indicated that this aptasensor has excellent selectivity.

3.6. Validation of the Aptasensor

Different SDZ concentrations (10, 20, and 30 ng/mL) were analyzed in spiked samples to demonstrate the properties of the aptasensor in highly complex biologic samples. The average recovery of these samples was between 87.19% and 109.26%, and the coefficients of variation were between 3.13% and 13.14% (Table 2). Furthermore, the sensor showed a positive relationship with its HPLC (Figure 7), demonstrating that it can reliably detect SDZ in animal-derived foods and the environment. Overall, the simple synthesis process and better burst performance of MnO_2 nanosheets make the developed FRET-based aptasensor simpler and less expensive, showing the promising potential of the method for application in real samples.

Table 2. Mean recoveries and coefficients of variation in spiked samples (n = 5).

Sample	Spiked (μg/kg)	This Work			HPLC		
		Found (μg/kg)	Recovery (%)	CV (%)	Found (μg/kg)	Recovery (%)	CV (%)
Soil	10	8.96	89.58	12.71	10.55	105.45	3.39
	20	20.47	102.36	8.03	19.27	96.36	9.51
	30	31.75	105.84	7.45	28.73	95.76	9.38
Lake water	10	8.72	87.19	11.21	9.45	94.55	8.33
	20	19.88	99.39	7.81	19.82	99.09	8.33
	30	31.27	104.23	9.98	30.55	101.82	4.67
River water	10	10.35	98.71	11.99	9.64	96.36	8.98
	20	21.33	106.65	9.28	18.18	90.91	9.60
	30	30.62	102.07	3.13	28.55	95.15	7.73
Egg	10	10.22	102.18	7.74	9.27	92.73	3.79
	20	20.90	104.50	7.66	18.00	90.00	5.33
	30	32.78	109.26	13.14	27.82	92.73	7.08
Beef	10	10.74	107.39	9.10	10.00	100.00	9.05
	20	20.85	104.23	7.16	19.64	98.18	6.39
	30	32.46	108.19	5.49	26.91	89.70	2.68

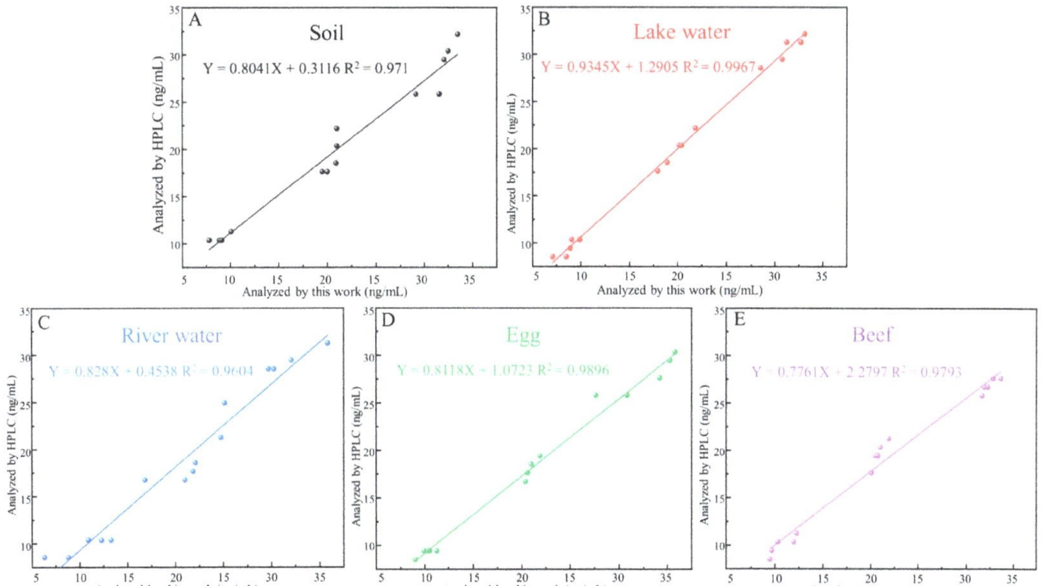

Figure 7. Standard curves corresponding to detection in food matrix, including (**A**) soil, (**B**) lake water, (**C**) river water, (**D**) egg, (**E**) beef.

4. Conclusions

In summary, we first developed a fluorescent aptasensor based on FAM-SDZ30-1 and MnO_2 to detect SDZ in food and the environment. The molecular dynamics simulations showed that RMSD values fluctuated from 10 to 100 ns around 0.6 nm in the SDZ30-1 and SDZ systems, indicating the high selectivity of SDZ30-1. The linear detection range (5–40 ng/mL) and LOD (3.25 ng/mL) of the aptasensor were sensitive to SDZ detection. The specificity of this aptasensor was satisfactory compared to various antibiotics. Additionally, the correlation between the aptasensor and HPLC was high ($R^2 \geq 0.9604$), demonstrating its precision in detecting SDZ. Therefore, this fluorescent aptasensor is simple to operate, low cost, and selective, and can be used as a reliable tool to monitor SDZ in different samples with promising applications in detection.

Author Contributions: X.Z.: main operator of experiment and writing—review. L.Y.: experimental second operator. Q.S.: analyze experimental data and investigation. L.Z.: methodology and writing—original draft. T.L.: writing—review and editing. All authors have read and agreed to the published version of the manuscript.

Funding: This research has been supported by the National Natural Science Foundation of China (Grant No. 31671939) and the Postgraduate Scientific Research and Innovation Project of Chongqing Municipal Education Commission (CYS22562).

Institutional Review Board Statement: Not applicable.

Informed Consent Statement: Not applicable.

Data Availability Statement: Not applicable.

Conflicts of Interest: The authors declare that there are no known competing financial interests or personal relationships that could have appeared to influence the work reported in this paper.

References

1. Shi, T.; Tan, L.; Fu, H.; Wang, J. Application of molecular imprinting polymer anchored on CdTe quantum dots for the detection of sulfadiazine in seawater. *Mar. Pollut. Bull.* **2019**, *146*, 591–597. [CrossRef]
2. Kokulnathan, T.; Kumar, E.A.; Wang, T.J.; Cheng, I.C. Strontium tungstate-modified disposable strip for electrochemical detection of sulfadiazine in environmental samples. *Ecotoxicol. Environ. Saf.* **2021**, *208*, 111516. [CrossRef] [PubMed]
3. Sakthivel, R.; Kubendhiran, S.; Chen, S.-M.; Chen, T.-W.; Al-Zaqri, N.; Alsalme, A.; Alharthi, F.A.; Abu Khanjer, M.M.; Tseng, T.-W.; Huang, C.-C. Exploring the promising potential of MoS_2-RuS_2 binary metal sulphide towards the electrocatalysis of antibiotic drug sulphadiazine. *Anal. Chim. Acta* **2019**, *1086*, 55–65. [CrossRef] [PubMed]
4. Ms, A.; Ep, B.; Hrn, C.; Aa, D.; Fs, E. Development of a rapid efficient solid-phase microextraction: An overhead rotating flat surface sorbent based 3-D graphene oxide/ lanthanum nanoparticles @ Ni foam for separation and determination of sulfonamides in animal-based food products. *Food Chem.* **2021**, *373 Pt A*, 131421. [CrossRef]
5. Dong, F.; Li, C.; Crittenden, J.; Zhang, T.; Lin, Q.; He, G.; Zhang, W.; Luo, J. Sulfadiazine destruction by chlorination in a pilot-scale water distribution system: Kinetics, pathway, and bacterial community structure. *J. Hazard. Mater.* **2019**, *366*, 88–97. [CrossRef]
6. Li, Z.B.; Cui, P.L.; Liu, J.; Liu, J.X.; Wang, J.P. Production of generic monoclonal antibody and development of chemiluminescence immunoassay for determination of 32 sulfonamides in chicken muscle. *Food Chem.* **2020**, *311*, 125966. [CrossRef]
7. Patyra, E.; Nebot, C.; Gavilán, R.E.; Cepeda, A.; Kwiatek, K. Development and validation of an LC-MS/MS method for the quantification of tiamulin, trimethoprim, tylosin, sulfadiazine and sulfamethazine in medicated feed. *Food Addit. Contam. Part A Chem. Anal. Control. Expo. Risk Assess.* **2018**, *35*, 882–891. [CrossRef]
8. Dil, E.A.; Ghaedi, M.; Mehrabi, F.; Tayebi, L. Highly selective magnetic dual template molecularly imprinted polymer for simultaneous enrichment of sulfadiazine and sulfathiazole from milk samples based on syringe-to-syringe magnetic solid-phase microextraction. *Talanta* **2021**, *232*, 122449. [CrossRef]
9. Errayess, S.A.; Lahcen, A.A.; Idrissi, L.; Marcoaldi, C.; Chiavarini, S.; Amine, A. A sensitive method for the determination of Sulfonamides in seawater samples by Solid Phase Extraction and UV-Visible spectrophotometry. *Spectrochim. Acta A Mol. Biomol. Spectrosc.* **2017**, *181*, 276–285. [CrossRef]
10. Zeng, Y.; Liang, D.; Zheng, P.; Zhang, Y.; Wang, Z.; Mari, G.M.; Jiang, H. A simple and rapid immunochromatography test based on readily available filter paper modified with chitosan to screen for 13 sulfonamides in milk. *J. Dairy Sci.* **2021**, *104*, 126–133. [CrossRef] [PubMed]
11. Moudgil, P.; Bedi, J.S.; Aulakh, R.S.; Gill, J.P.S.; Kumar, A. Validation of HPLC Multi-residue Method for Determination of Fluoroquinolones, Tetracycline, Sulphonamides and Chloramphenicol Residues in Bovine Milk. *Food Anal. Methods* **2019**, *12*, 338–346. [CrossRef]
12. Wang, D.; Zhang, J.; Huang, Z.; Yang, Y.; Fu, T.; Yang, Y.; Lyu, Y.; Jiang, J.; Qiu, L.; Cao, Z. Robust Covalent Aptamer Strategy Enables Sensitive Detection and Enhanced Inhibition of SARS-CoV-2 Proteins. *ACS Cent. Sci.* **2023**, *9*, 72–83. [CrossRef]
13. Karadjian, G. Aptamer-Based Technologies for Parasite Detection. *Sensors* **2023**, *23*, 562. [CrossRef]
14. Sun, S. The Research Advances of Aptamers in Hematologic Malignancies. *Cancers* **2023**, *15*, 300. [CrossRef]
15. Zhu, Y.; Wang, J.; Xie, H.; Fu, C.; Zhou, J.; Liu, H.; Zeng, P.; Sun, Y. Double Signal Amplification Strategy for Dual-Analyte Fluorescent Aptasensors for Visualizing Cancer Biomarker Proteins. *Anal. Chem.* **2022**, *94*, 10451–10461. [CrossRef]

16. Wang, X.; Rong, X.; Zhang, Y.; Luo, F.; Qiu, B.; Wang, J.; Lin, Z. Homogeneous Photoelectrochemical Aptasensors for Tetracycline Based on Sulfur-Doped g-C3N4/n-GaN Heterostructures Formed through Self-Assembly. *Anal. Chem.* **2022**, *94*, 3735–3742. [CrossRef]
17. Lee, E.S.; Kim, E.J.; Park, T.K.; Bae, D.W.; Cha, S.S.; Kim, T.W.; Kim, Y.P. Gold nanoparticle-assisted SELEX as a visual monitoring platform for the development of small molecule-binding DNA aptasensors. *Biosens. Bioelectron.* **2021**, *191*, 113468. [CrossRef]
18. Yang, H.; Xu, D. Highly-sensitive and simple fluorescent aptasensor for 17b-estradiol detection coupled with HCR-HRP structure. *Talanta* **2022**, *240*, 123094. [CrossRef] [PubMed]
19. Sameiyan, E.; Khoshbin, Z.; Lavaee, P.; Ramezani, M.; Alibolandi, M.; Abnous, K.; Taghdisi, S. A bivalent binding aptamer-cDNA on MoS_2 nanosheets based fluorescent aptasensor for detection of aflatoxin M1. *Talanta* **2021**, *235*, 122779. [CrossRef]
20. Yan, X.; Wang, Y.; Kou, Q.; Sun, Q.; Tang, J.; Yang, L.; Chen, X.; Xu, W.; Le, T. A novel aptasensor based on $Fe_3O_4/Au/g\text{-}C_3N_4$ for sensitive detection of sulfameter in food matrices. *Sens. Actuators B Chem.* **2021**, *353*, 131148. [CrossRef]
21. Wang, X.; Zhang, L.; Hao, A.; Shi, Z.; Huang, H. Silica-Coated Silver Nanoparticles Decorated with Fluorescent CdTe Quantum Dots and DNA Aptamers for Detection of Tetracycline. *ACS Appl. Nano Mater.* **2020**, *3*, 9796–9803. [CrossRef]
22. Yan, X.; Song, Y.; Zhu, C.; Song, J.; Du, D.; Su, X.; Lin, Y. Graphene Quantum Dot-MnO_2 Nanosheet-Based Optical Sensing Platform: A Sensitive Fluorescence "Turn Off-On" Nanosensor for Glutathione Detection and Intracellular Imaging. *ACS Appl. Mater. Interfaces* **2016**, *8*, 21990–21996. [CrossRef] [PubMed]
23. Hu, C.; Kong, X.J.; Yu, R.Q.; Chen, T.T.; Chu, X. MnO_2 Nanosheet-based Fluorescence Sensing Platform for Sensitive Detection of Endonuclease. *Anal. Sci.* **2017**, *33*, 783–788. [CrossRef]
24. Xu, M.; Zhuang, J.; Jiang, X.; Liu, X.; Tang, D. A three-dimensional DNA walker amplified FRET sensor for detection of telomerase activity based on the MnO_2 nanosheet-upconversion nanoparticle sensing platform. *Chem. Commun.* **2019**, *55*, 9857–9860. [CrossRef]
25. Ouyang, Q.; Wang, L.; Ahmad, W.; Rong, Y.; Li, H.; Hu, Y.; Chen, Q. A highly sensitive detection of carbendazim pesticide in food based on the upconversion-MnO(2) luminescent resonance energy transfer biosensor. *Food Chem.* **2021**, *349*, 129157. [CrossRef]
26. Li, J.; Du, B.; Li, Y.; Wang, Y.; Wu, D.; Wei, Q. A turn-on fluorescent sensor for highly sensitive mercury(ii) detection based on a carbon dot-labeled oligodeoxyribonucleotide and MnO_2 nanosheets. *New J. Chem.* **2018**, *42*, 1228–1234. [CrossRef]
27. Paissoni, C.; Spiliotopoulos, D.; Musco, G.; Spitaleri, A. GMXPBSA 2.0: A GROMACS tool to perform MM/PBSA and computational alanine scanning. *Comput. Phys. Commun.* **2014**, *185*, 105–107. [CrossRef]
28. Elkarhat, Z.; Charoute, H.; Elkhattabi, L.; Barakat, A.; Rouba, H. Potential inhibitors of SARS-CoV-2 RNA dependent RNA polymerase protein: Molecular docking, molecular dynamics simulations and MM-PBSA analyses. *J. Biomol. Struct. Dyn.* **2020**, *40*, 361–374. [CrossRef]
29. Yan, X.; Yang, L.; Tang, J.; Wen, X.; Chen, X.; Zheng, X.; Chen, L.; Li, J.; Le, T. High-Sensitive FAM Labeled Aptasensor Based on $Fe_3O_4/Au/g\text{-}C_3N_4$ for the Detection of Sulfamethazine in Food Matrix. *Biosensors* **2022**, *12*, 759. [CrossRef]
30. Chen, X.; Yang, L.; Tang, J.; Wen, X.; Zheng, X.; Chen, L.; Li, J.; Xie, Y.; Le, T. An AuNPs-Based Fluorescent Sensor with Truncated Aptamer for Detection of Sulfaquinoxaline in Water. *Biosensors* **2022**, *12*, 513. [CrossRef]
31. Yang, L.; Chen, X.; Wen, X.; Tang, J.; Zheng, X.; Li, J.; Chen, L.; Jiang, S.; Le, T. A label-free dual-modal aptasensor for colorimetric and fluorescent detection of sulfadiazine. *J. Mater. Chem. B* **2022**, *10*, 6187–6193. [CrossRef] [PubMed]
32. Tang, J.; Kou, Q.; Chen, X.; Wang, Y.; Yang, L.; Wen, X.; Zheng, X.; Yan, X.; Le, T. A novel fluorescent aptasensor based on mesoporous silica nanoparticles for the selective detection of sulfadiazine in edible tissue. *Arab. J. Chem.* **2022**, *15*, 104067. [CrossRef]

Article

Biosensing Platform for the Detection of Biomarkers for ALI/ARDS in Bronchoalveolar Lavage Fluid of LPS Mice Model

Nuha Khalid Alekhmimi [1,2], Dana Cialla-May [2,3], Qasem Ramadan [1], Shimaa Eissa [4,5], Jürgen Popp [2,3], Khaled Al-Kattan [6] and Mohammed Zourob [1,*]

1 Department of Chemistry, Alfaisal University, Al Zahrawi Street, Al Maather, AlTakhassusi Rd, Riyadh 11533, Saudi Arabia
2 Leibniz Institute of Photonic Technology, Albert-Einstein-Straße 9, 07745 Jena, Germany
3 Institute of Physical Chemistry (IPC) and Abbe Center of Photonics (ACP), Friedrich Schiller University Jena, Helmholtzweg 4, 07743 Jena, Germany
4 Department of Chemistry, Khalifa University of Science and Technology, Abu Dhabi P.O. Box 127788, United Arab Emirates
5 Advanced Materials Chemistry Center (AMCC), Khalifa University of Science and Technology, Abu Dhabi P.O. Box 127788, United Arab Emirates
6 College of Medicine, Alfaisal University, Al Zahrawi Street, Al Maather, Al Takhassusi Rd, Riyadh 11533, Saudi Arabia
* Correspondence: mzourob@alfaisal.edu

Abstract: Acute respiratory distress syndrome (ARDS) is a worldwide health concern. The pathophysiological features of ALI/ARDS include a pulmonary immunological response. The development of a rapid and low-cost biosensing platform for the detection of ARDS is urgently needed. In this study, we report the development of a paper-based multiplexed sensing platform to detect human NE, PR3 and MMP-2 proteases. Through monitoring the three proteases in infected mice after the intra-nasal administration of LPS, we showed that these proteases played an essential role in ALI/ARDS. The paper-based sensor utilized a colorimetric detection approach based on the cleavage of peptide–magnetic nanoparticle conjugates, which led to a change in the gold nanoparticle-modified paper sensor. The multiplexing of human NE, PR3 and MMP-2 proteases was tested and compared after 30 min, 2 h, 4 h and 24 h of LPS administration. The multiplexing platform of the three analytes led to relatively marked peptide cleavage occurring only after 30 min and 24 h. The results demonstrated that MMP-2, PR3 and human NE can provide a promising biosensing platform for ALI/ARDS in infected mice at different stages. MMP-2 was detected at all stages (30 min–24 h); however, the detection of human NE and PR3 can be useful for early- (30 min) and late-stage (24 h) detection of ALI/ARDS. Further studies are necessary to apply these potential diagnostic biosensing platforms to detect ARDS in patients.

Keywords: ALI/ARDS; biosensors; PR3; human NE; MMP-2

Citation: Alekhmimi, N.K.; Cialla-May, D.; Ramadan, Q.; Eissa, S.; Popp, J.; Al-Kattan, K.; Zourob, M. Biosensing Platform for the Detection of Biomarkers for ALI/ARDS in Bronchoalveolar Lavage Fluid of LPS Mice Model. *Biosensors* **2023**, *13*, 676. https://doi.org/10.3390/bios13070676

Received: 30 March 2023
Revised: 6 June 2023
Accepted: 17 June 2023
Published: 25 June 2023

1. Introduction

Acute respiratory distress syndrome (ARDS) is a life-threatening lung injury due to positive end-expiratory pressure resulting from the inflammatory response of the innate immune system. Acute lung injury (ALI) is a mild form of ARDS, with a ratio of arterial oxygen to inspired oxygen (PaO_2/FIO_2) of less than 300 mm Hg. This definition, which was updated in 2012 by the ARDS Definition Task Force, an international panel of experts, is known as the Berlin Definition. The Berlin Definition proposed three distinct classifications of ARDS based on the degree of hypoxemia: mild (200 mm Hg < $PaO_2/FiO_2 \leq$ 300 mm Hg), moderate (100 mm Hg < $PaO_2/FiO_2 \leq$ 200 mm Hg) and severe (PaO_2/FiO_2 $\leq$ 100 mm Hg) [1]. Despite removing the term "ALI" from the Berlin Definition, it is still used in research and clinical arenas [2].

ALI is considered an innate immune inflammatory response that can develop into a major clinical syndrome, ARDS, as well as interstitial edema and impaired gas exchange. Pulmonary neutrophil infiltration is a common pathophysiological characteristic of ALI/ARDS. Polymorphonuclear leukocytes (PMN) and pulmonary dysfunction are integral components of the alveolar infiltrate, which are mainly attributed to collateral damage induced by PMN invasion [3]. Lung injury and the infection-induced upregulation of proinflammatory cytokines (e.g., TNF-α and IL-8) and chemokines are remarkable manifestations of alveolar epithelial cells and the macrophage immune response [3]. This leads to neutrophil infiltration due to acute inflammatory disorders caused by proteases. Recent studies have shown that the neutrophil secretion of granule proteins leads to targeting and cleaving bacterial virulence factors [4]. It was found that neutrophil serine proteinases cleave bacteria in both humans and mice [5]. Proteinase 3 (PR3) can convert TNF-α cytokines and proceed to degrade membrane-bound pro-TNF-α cytokines, either by membrane-bound matrix metalloproteinase or a TNF-α converting enzyme, during the inflammation process [6].

ALI/ARDS is associated with high morbidity and mortality rates, with a wide variation in the reported population incidence, making it a global health problem. Despite improvements in treatment regimens in intensive care units, ARDS mortality has remained at around 40% for the last several years [7,8]. Therefore, the early diagnosis of ALI/ARDS cases is essential.

When a lung injury occurs, blood-circulating neutrophils migrate to the injured site and attach to the endothelium, releasing oxygen radicals and their internal constituents, which are stored in different granules [9]. The azurophil granules, for instance, consist of serine proteases, neutrophil elastase (NE), cathepsin G and PR3. Their main function is to release their granular content into the extracellular space via exocytosis [10], and break down the pathogens in phagosomes [6]. It is well known that the uncontrolled production of the serine protease, which PMNs produce, is associated with various inflammatory disease states, lung injuries and sepsis.

The PR3 protease is one of the main serine proteases that is secreted from neutrophils. It upregulates inflammation cascades [11,12] and induces PR3 lung damage. One of PR3's functions is to initiate neutrophil apoptosis, which has various implications in autoimmune disease [13]. In early studies, PR3 was considered an enzyme that degrades elastin and causes pulmonary emphysema after intra-tracheal instillation in hamsters [14]. PR3 is also known to assign the same conserved catalytic cleft with 35% homology to cathepsin G [15] and 39% to neutrophil serine proteases 4 [16]. Moreover, neutrophil's serine protease granule has a dynamic range that indicates PR3 when it rises to 13.4 mM [17]. Boxio et al. have shown that PR3 is a potent enzyme that can degrade E-cad, characterized as one of the most important, ubiquitous adherent junction proteins, and E-cad, a glycoprotein member of the cadherin superfamily that only overlaps the membrane once [18].

Matrix metalloproteinases (MMPs) are zinc-dependent proteinases capable of cleaving extracellular matrix proteins and enzymes, which have an important role in lung dysfunction [19–22]. MMPs modify cell–matrix and cell–cell interactions by their ability to cleave proteins such as collagen from large components to small proinflammatory mediators [23]. The activation of MMP-2 and MMP-9 by LPS induces the release of TNF-α, which can raise MMP-2 and MMP-9 levels in the blood of patients with sepsis in response to a Gram-negative bacterial infection. MMP-2/MMP-9 has been correlated with the severity of sepsis [24].

Various methods, such as CT scans [25–28] chest X-rays and MRI, are currently used for the diagnosis of ALI/ARDS [14,15]. However, using these methods requires well-trained laboratory personnel and costly instruments [28], and may increase lung injury risk [29–31]. The ALI "sniffer," which can identify ALI with a sensitivity of 96% (95% CI 94–98) and a specificity of 89% (95% CI 88–90), can accurately determine if critically ill patients will develop ALI syndrome [32]. However, it is costly, and is only available in central healthcare centers and laboratories. The enzyme-linked immunosorbent assay (ELISA) and other immunoassays have also been employed to detect ALI/ARDS biomarkers in serum, plasma and bronchoalveolar lavage (BAL) [33,34], yet the ELISA is a time-consuming multi-step assay that requires costly reagents and well-trained personnel in centralized labs [35,36]. Therefore, developing rapid, low-cost ALI/ARDS methods is paramount.

The use of biosensors is an excellent alternative to traditional assays, in which the on-site monitoring of multiple biomarkers can be performed for the diagnosis of diseases. Few biosensors have been reported for the detection of PR3, MMP-2 and human NE (NE). A recent study by Griffith et al. identified the epitopes on a PR3-linked catalytic site using a resonant mirror biosensor for different patients with Wegener's granulomatosis [19]. A computer-generated model showed that four out of five epitopes were quietly connected with the catalytic site and could form a contourable epitope. Wang et al. [25] developed an electrochemical peptide cleavage-based biosensor for MMP-2 detection with exonuclease III-assisted cycling signal amplification that could detect MMP-2 at concentrations as low as 0.15 pg/mL, with a dynamic range of 0.5 pg/mL–50 ng/mL. A recent study reported the detection of leukemia using electrochemical cytosensing to detect MMP-2 in human serum [26]. A colorimetric biosensor for the detection of NE has been reported for the diagnosis of periodontitis [27]. The absence of rapid, low-cost biosensors for detecting lung injuries, identifying and monitoring therapy, and predicting the prognosis of ARDS necessitates the availability of such diagnostic tools to detect ALI at an early stage before it develops into ARDS.

To the best of our knowledge, the diagnosis of ALI/ARDS through the detection of PR3, MMP-2 and NE as biomarkers has not been reported previously. In this study, we developed and characterized a paper-based multiplexing sensing method for detecting ALI/ARDS. Colorimetric paper-based biosensors are a simple detection method at ALI's early and late stages. While paper has been used as a substrate for chemical analyses for many years, the concept of paper-based magnetic nanoparticles has only recently been introduced [36,37]. The easy synthesis of gold nanoparticles has led to increasing interest in applying AuNPs in interfacing biological recognition in several biosensor applications [38].

Specific peptide sequences for PR3, MMP-2 and NEs were conjugated with magnetic nanobeads, and a colorimetric probe was used to detect the cleavage of the peptides for the detection of ALI/ARDS in infected mice. The early detection of ALI/ARDS may enable an understanding of the mechanisms underlying pulmonary therapy for patients with ALI/ARDS, thus reducing the risk of life-threatening lung injuries.

2. Materials and Methods

2.1. Materials

The Gainland Chemical Company (Sandycroft, Deeside, UK) provided a Diff-Quick stain. The TNF-α and MIP-2 [39–43] kits were purchased from R&D System (Minneapolis, MN, USA). All of the reagents and solutions were prepared using deionized water and stored at −20 °C until their usage.

The MMP-2, PR3, NE peptides and lipopolysaccharide (LPS) *E. coli* 026:B6 [44–54] were synthesized by Pepmic Co. (Suzhou, Jiangsu, China). The PR3 peptide used was GIATDCMLMPEQ [55], the MMP-2 peptide was GPQVLANIPHFP [56], and NE peptide was APPEEIMDRQ [57]. Sodium phosphate and monopotassium phosphate, tris-base, sodium acetate, lipopolysaccharide (LPS) *E. coli* 026:B6, bovine serum albumin (BSA), sodium chloride, phosphate-buffered saline (PBS), ethyl-3-(3-dimethylaminopropyl)-carbodiimide (EDC), N-hydroxysuccinimide (NHS) and phosphate-buffered ethylenediaminetetraacetic acid (EDTA) were purchased from Sigma-Aldrich (Darmstadt, Germany).

2.2. Experimental Methods

2.2.1. Mice Preparation

Wild-type C57BL/6J mice were purchased from Jackson Laboratories (Bar Harbor, ME, USA) at 6–10 weeks of age and 16–22 gms in weight. The mice were housed in the Department of Comparative Medicine at the King Faisal Specialist Hospital and Research Center (KFSH&RC) under specific pathogen-free conditions with a 12-h light/dark cycle and free access to food and water. The mice were used for in vivo experiments and as a source for collecting BAL fluid. The experimental protocols were revised and approved by the KFSH&RC Animal Care and Use Committee (RAC# 1MED 1672).

2.2.2. LPS Model for ALI/ARDS

LPS from *E. coli* 026: B6 was used as an inhalation model for ALI/ARDS [41–43]. The LPS (50 µg in 50 µL normal saline) was administered intra-nasally by applying drops from a pipette placed on the nares. Each drop was allowed to be inhaled before the next drop was administered 1–2 min later. As a control, mice were given 50 µL of normal saline intra-nasally. The animal study was conducted according to a protocol approved by the KFSH&RC Animal Care and Use Committee (RAC# 1MED 1672).

2.2.3. Bronchoalveolar Lavage Fluid (BALF)

After 30 min, 2 h, 4 h and 24 h of LPS treatment, the mice were anesthetized by the intraperitoneal injection of a mixer of ketamine (80 mg/kg) and xylazine (5 mg/kg). An incision was made in the neck, and the trachea was punctured with a 1 cc needle. Subsequently, the trachea was cannulated with a 24-gauge catheter and secured with 40 sutures. The lungs were lavaged with 6–8 aliquots of 0.6 mL of sterile PBS containing 0.1% EDTA. The BAL mice samples were stored at −80 °C until they were used again.

2.2.4. Cytospin and Cell Count Differentials

The BAL fluid collected from the untreated and LPS-treated mice was centrifuged at 300× *g* for 10 min at 4 °C. The pellet cells were re-suspended in 5.0 mL of RPMI fresh media containing 10% fetal bovine serum and 5% penicillin/streptomycin. Subsequently, 200 µL was added to a glass slide and centrifuged using cytospin at 800 rpm for 3 min to allow the cells to adhere. The slides were left to dry overnight and stained with a Diff-Quick stain kit according to the manufacturer's instructions. After drying, the slides were briefly placed in an eosin solution for 30 s, and in a methylene blue solution for 30 s. Then, they were washed with tap water to remove the residual dye. Next, the coverslips were placed on the slides. The AMΦ and neutrophils were identified morphologically and counted by two blinded observers (data not shown).

2.2.5. Cytokines and Chemokine Measurements with ELISA

TNF-α and MIP-2 ELISA kits were used to assess AMΦ secretions according to the manufacturer's instructions. BAL fluid was collected from the mice 4 h after LPS infection. ELISA kits were used to determine the anti-mouse TNF-α and MIP-2. After adding the substrate streptavidin and the termination of a 30-min reaction, the product was assessed calorimetrically with an ELISA reader. The results were compared to the TNF-α and MIP-2 standard curves described in the kits' manuals.

2.2.6. Conjugation of PR3, MMP-2 and Human NE Peptides with Magnetic Beads

A total of 100 µL of 1 mg/mL aliquots of peptide substrates were dissolved in dimethyl sulfoxide, then mixed with pre-washed magnetic beads and a freshly prepared EDC/NHS coupling solution. The mixture was rotated at room temperature and incubated overnight at 4 °C. The uncoupled peptides were removed by washing the beads 2–3 times with a washing buffer. The bead-activated peptides of PR3, MMP-2 and human NE were stored at 4 °C until further use.

2.2.7. Sensing Platform Fabrication

Self-adhesive sheets were purchased from Whatman (London, UK); they were coated with a gold layer with a thickness of 30 nm using a sputtering machine in a clean room at the King Abdullah University of Science and Technology (KAUST). The gold-coated sheet was cut into 1–2 mm sections and mounted over a plastic strip at a separation distance of

3 mm. The plastic strip was used as a physical support for the bio-functionalization process, as shown in Scheme 1A.

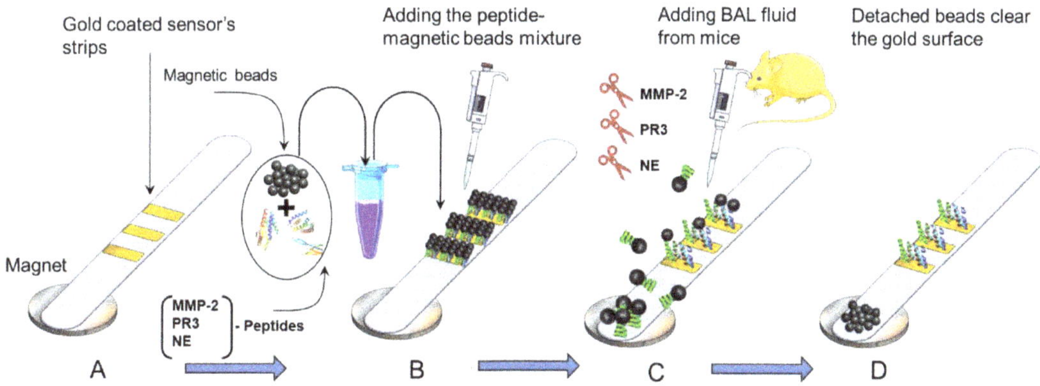

Scheme 1. Schematic representation of the working principle of the paper-based ALI/ARDS sensor: (**A**) gold-coated adhesive tape was stacked over a paper substrate, (**B**) peptide-conjugated magnetic beads were added to the gold surface to form a functional sensor, (**C**) the LPS-treated BAL sample from the mice was added, and (**D**) uncovering the gold-colored surface due to protease peptide cleavage.

A layer of peptide–magnetic nanoparticle (MNPs) conjugates was immobilized onto the gold surface and allowed to dry at room temperature for 30 min (Scheme 1A). Subsequently, a permanent external magnet with dimensions of 10 mm × 10 mm × 5 mm and a magnetic flux density of 3300 Gauss at a 2 mm distance was passed over the functionalized strip to remove any unattached MNPs. Upon immobilization, the sensor's gold surface was masked and turned black (Scheme 1B). After that, a round paper magnet was positioned 2–3 mm below the sensor strip. When the protease's cleavage occurred, the black surface turned gold again due to the cleavage of the protease peptide that occurred when adding the sample containing the analytes, as shown in Scheme 1C.

2.2.8. Quantitative Measurements of the Paper-Based Biosensing Platform

The images of the sensors were taken and saved in JPEG format, and processed using ImageJ software to calculate the quantitative data. The concentrations of the proteases of the paper-based biosensors and the multiplex data were calculated by dividing the cleaved area (yellow) by the total black sensor area. Quantitation experiments were conducted using different concentrations of proteases. The experiments were conducted in triplicate.

2.2.9. In Vitro Testing Using FRET Assay (Fluorescence Resonance Energy Transfer Sensors)

The three FRET peptide substrates used in this study were purchased from Pepmic (Suzhou, China), with a purity greater than 95%. All of the peptides were dissolved in 100% dimethyl sulfoxide and stored at −80 °C. The proteolytic activities of the different substrates were individually monitored using 0.5 µL of 800 µM of FAM/DABCYL-modified peptides in 100 µL of MMP-2, PR3 and NE protease solutions in PBS buffer. BAL mice samples treated with normal saline were used.

The change in the fluorescence intensity of the FAM/DABCYL-modified peptides was monitored every minute for 2 h at 37 °C, at excitation and emission wavelengths of 485 and 535 nm, respectively. The comparative fluorescence intensity for the various protease probes in the BAL mice samples was analyzed by plotting the mean fluorescence intensity against time for each peptide. Normal saline was used in the blanks.

2.2.10. Statistical Analysis

A *t*-test was used to compare the two means. An ANOVA test (San Diego, CA, USA) was used for multiple comparisons, followed by a *t*-test with Bonferroni corrections in Graphpad (San Diego, CA, USA).

3. Results and Discussion

Pre-clinical and clinical studies have indicated that the inflammatory response to direct and indirect insults to the lung plays a pivotal role in the pathogenesis of ALI/ARDS [2]. The early detection of bacterial infection can prevent the spread of ALI/ARDS and its progression. In this research, we developed a multiplexed paper-based sensor for the simultaneous detection of MMP-2, PR3 and NE proteases in an LPS mice model (mimic the ALI/ARDS). using specific peptides for the three proteases. The peptides were immobilized individually onto AuNP-modified paper-based surfaces (Scheme 1).

3.1. Assessing ALI/ARDS by Evaluating TNF-α/MIP-2 in a BAL Secretion after LPS Stimulation in Mice

The lung inflammation-associated upregulation of TNF-α and MIP-2 has been reported in several studies [38–40]. The LPS-intra-nasal injection of animal models has emphasized the role of TNF-α and MIP-2 secretion in ALI/ARDS [41,42]. An ELISA was used to measure the levels of LPS-induced prototypic cytokines, TNF-α and MIP-2. Figure 1 shows the comparison of control animals treated with normal saline (NS); TNF-α and MIP-2 secretions significantly increased (10-fold) in BAL fluids in response to LPS stimulation. The relative increments of the MIP-2 released were found to be higher than those of TNF-α in mice with ALI/ARDS, $p < 0.05$ (Figure 1). This finding, which agrees with those of previous studies [21,38,41,43], demonstrated that the TNF-α cytokine and MIP-2 chemokine play a significant role in upregulating lung inflammation in mice.

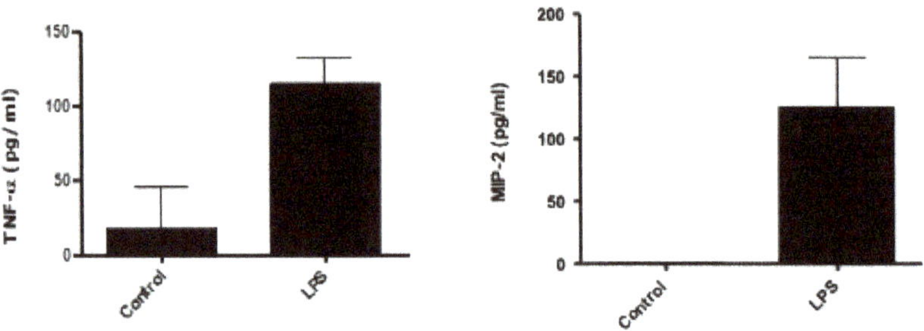

Figure 1. TNF-α and MIP-2 levels in BAL collected from post-LPS-treated mice (4 h) compared to normal saline (NS = control). TNF-α and MIP-2 significantly increased (10-fold) in BAL fluids, n = 3, $p < 0.05$.

3.2. Quantitative Measurements for Various Proteases Using the Paper-Based Assay

The sensors for the three proteases were prepared using the method reported previously [27,58,59]. Briefly, three specific peptides for PR3, MMP-2 and NE were attached covalently to magnetic nanobeads via the reaction of the peptide N terminal to the carboxyl-modified magnetic beads. The C-terminal of the peptide attached to the gold nanoparticle coated layer placed on a paper substrate via self-assembly. The attachment of the peptide–magnetic particle conjugates to the gold surface changes the color of the gold from yellow to black. On the other hand, the binding of the protease in the sample with the sensor surface leads to the cleavage of the peptide and the detachment of the magnetic particles from the sensor surface, which can then be removed using a magnetic placed underneath the paper. The sensors were tested to quantitatively detect the proteolytic activity of each protease (PR3, MMP-2 and NE). Upon proteolysis, the gold-colored sensor surface was revealed, leading to a significant change in the color, from black to yellow, which is visible to the naked eye. ImageJ software was used to quantitatively measure the current change upon peptide cleavage after binding with different concentrations of the three proteases. The peptide cleavage was correlated with the intensity of the revealed yellow color. Figure 2 shows that a gradual increase in the visible bare gold area was attained by increasing protease concentrations. It was observed that a high protease enzyme concentration allowed for faster peptide magnetic particles dissociation than low concentrations. As a negative control, normal saline showed no color change for the various proteases' sensors (Figure 2).

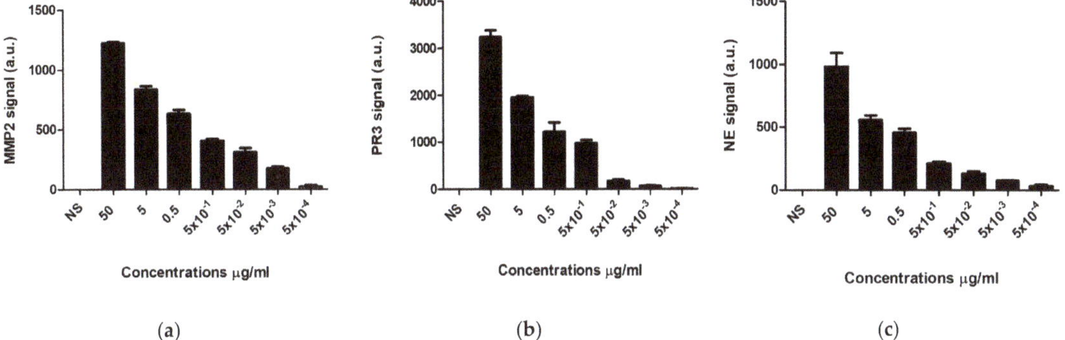

Figure 2. The response of protease sensors in normal saline (NS) and 50 µg/mL, 5 µg/mL, 0.5 µg/mL, 0.05 µg/mL, 50 ng/mL and 0.5 ng/mL for (**a**) MMP-2, (**b**) PR3 and (**c**) NE.

The colorimetric biosensors enabled a limit of detection for as low as 0.5 ng/mL within one min, which was determined by identifying the lowest protease concentration capable of cleaving the covalently attached peptide black magnetic nanobeads, which revealed the sensors' gold surface areas.

3.3. Measurements of Various Proteases in ALI/ARDS-Infected Mice

Figure 3 shows the time-dependent concentration profiles of the secreted PR3, MMP-2 and NE in the BAL samples from infected mice taken at different time intervals (30 min, 2 h, 4 h and 24 h) after treatment with the LPS. Figure 3a shows that the concentration of the MMP-2 protease increased 30 min and 2 h after LPS administration, and significantly increased at 4 h and 24 h, implying that the MMP-2 peptide could serve as a biomarker for detecting early inflammation and late-stage ALI/ARDS. These results confirm those of previous studies [23,41–53]. There was an observable high cleavage of PR3 peptide with the addition of BAL mice samples treated with LPS for 30 min, 2 h and 24 h (Figure 3b). However, no peptide cleavage was observed after 4 h of LPS administration, which indicates that the intra-nasal administration of LPS resulted in a cell count differential increase in PMN in the BAL fluid (data not shown). Therefore, these data suggest that PR3 can be used for the early detection (30 min–2 h) of early-stage ALI/ARDS and severe ARDS (after 24 h). Figure 3b showed significant increases in PR3 with time, except for the drop shown after 4 h.

NE plays an essential role in severe acute lung injuries. Elastase enzymes are secreted in ALI/ARDS during the inflammation process. Figure 3c shows a clear increase in NE peptide when exposed to BAL collected from LPS-treated mice samples after 4 and 24 h. Slight cleavage was detected at 30 min, but no cleavage was detected at 2 h. These results established that the NE biomarker could be useful for detecting late-stage ALI/ARDS. It is worth mentioning that the specificity of the biosensor was tested using nonspecific proteases and no cross reactivity was detected, indicating the high specificity of this assay.

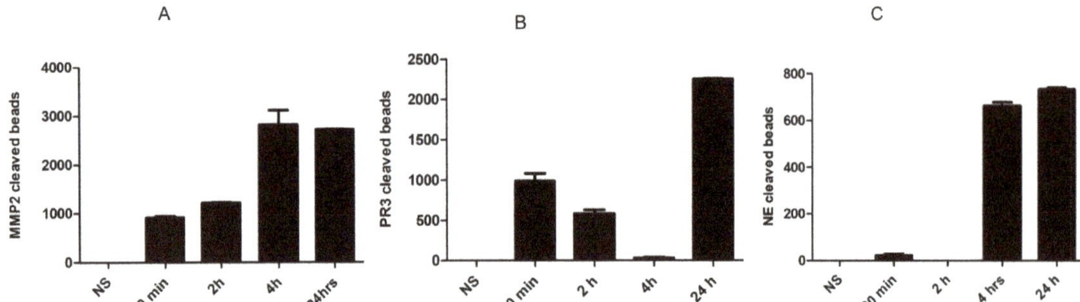

Figure 3. The profile of various proteases using the paper-based sensors after exposure to BAL collected from LPS-treated mice at different intervals. (**A**) MMP-2, (**B**) PR3 and (**C**) NE. Data were extracted using Image J software (n = 3).

3.4. Comparison of the Paper-Based Sensors with Fluorescence Resonance Energy Transfer (FRET) Sensors

FRET sensors have recently been used to quantitatively detect protease activity [55]. The results of the peptide-conjugated magnetic nanobeads with paper-based sensors were compared with peptides containing fluorescence and a quencher in a solution. The peptides were labeled with FAM fluorophore on the N-terminal and quencher DABCYL on the C-terminal. The FRET-specific peptides for the various proteases were incubated with different sets of BAL samples collected at different times following the introduction of LPS. Figure 4 shows the FRET results for the proteases MMP-2, PR3 and NE upon exposure to LPS samples collected at different times (0, 30 min, 2 h, 4 h and 24 h). A significant increase in fluorescence intensity was observed at 30 min, 2 h, 4 h and 24 h following ALI/ARDS, but no increase was seen in the BAL sample treated with NS. These results were in agreement with those of MMP-2 paper-based sensing cleavage, indicating that the colorimetric paper-based sensor can be used as a rapid detection tool. Meanwhile, PR3 FRET peptide sequences only led to a slight increase in the intensity of the fluorescence at 30 min and 2 h, but a significant increase after 24 h when exposed to the LPS-treated animal BAL samples (Figure 4b). Figure 4c also indicates a significant increase in fluorescence after 4 and 24 h after LPS administration. These results agree with those obtained using paper-based sensors. Eventually, the signal intensity profiles for different peptides at various times could detect ARDS in the early, severe stages of lung injuries in mice.

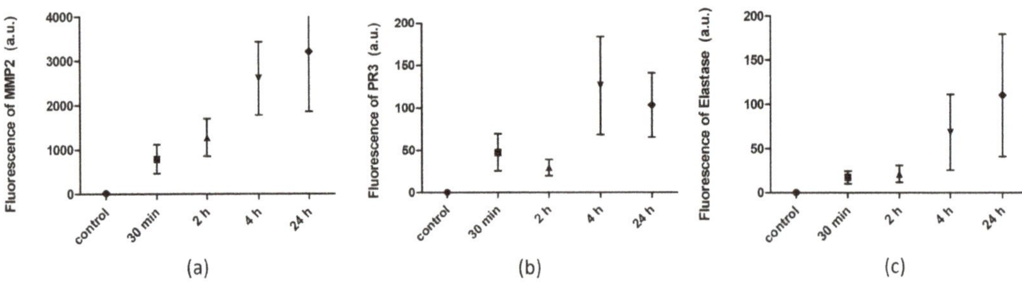

Figure 4. Comparison of FRET-based peptide substrate and paper-based biosensors with (**a**) MMP-2; (**b**) PR3; (**c**) NE (p value = 0.001, n = 4).

4. Conclusions

This study used proteases released from BAL in LPS-treated mice as biomarkers for the diagnosis of ALI/ARDS. Paper-based sensors provided a simple yet sensitive and rapid method of detection that can be read visually without laboratory-based equipment. The cleavage of the peptide–AuNPs conjugate, which was related to the protease concentration, resulted in exposing the sensor surface and revealing a gold color from a surface that was initially covered with black beads. The sensor showed good sensitivity, with a limit of detection of 5×10^{-4} µg/mL and a high specificity, as demonstrated by the multiplexed detection of three different proteases on the same paper sensor. We found that MMP-2 was the best biomarker for ALI/ARDS, and was able to detect the disease at all stages; thus, this multiplexing platform could provide very useful information for patients with ALI/ARDS. The reported multiplexed paper-based assay can provide rapid diagnosis of lung inflammation, showing potential advantages over existing traditional methods in terms of its cost reductions, simplicity and fast response.

Author Contributions: N.K.A. conducted all the experimental work, wrote first draft of the manuscript. D.C.-M. proof read the manuscript, Q.R. and S.E. rewrote the manuscript to be publishable. J.P., K.A.-K. and M.Z. supervised the work and secured the funding. All authors have read and agreed to the published version of the manuscript.

Funding: The authors extend their appreciation to the Deputyship for Research & Innovation, Ministry of Education in Saudi Arabia for funding this research work through Project No. 492.

Institutional Review Board Statement: The study was conducted in accordance with the rules and regulations of the government of Saudi Arabia, the KFMC/IRB polices and procedures, and the ICH good clinical practice guidelines under the IRB number IRB00010471.

Informed Consent Statement: Not applicable.

Data Availability Statement: Not applicable.

Acknowledgments: The authors would like to acknowledge Alfaisal University and the Al Ageel fund for their financial support for the conference. We would like to thank Falah Almohanna, Peter Kvietys and the team at the Comparative Medicine at KFSHRC for their technical assistance in the in vivo experiments.

Conflicts of Interest: The authors declare that they have no known competing financial interests or personal relationships that could have appeared to influence the research reported in this study.

References

1. Force, A.D.; Ranieri, V.M.; Rubenfeld, G.D.; Thompson, B.; Ferguson, N.; Caldwell, E.; Fan, E.; Camporota, L.; Slutsky, A.S. Acute Respiratory Distress Syndrome. *JAMA* **2012**, *307*, 2526–2533.
2. Butt, Y.; Kurdowska, A.; Allen, T.C. Acute Lung Injury: A Clinical and Molecular Review. *Arch. Pathol. Lab. Med.* **2016**, *140*, 345–350. [CrossRef]
3. Grommes, J.; Soehnlein, O. Contribution of Neutrophils to Acute Lung Injury. *Mol. Med.* **2011**, *17*, 293. [CrossRef]
4. Weinrauch, Y.; Drujan, D.; Shapiro, S.D.; Weiss, J.; Zychlinsky, A. Neutrophil Elastase Targets Virulence Factors of Enterobacteria. *Nature* **2002**, *417*, 91. [CrossRef]
5. López-Boado, Y.S.; Espinola, M.; Bahr, S.; Belaaouaj, A. Neutrophil Serine Proteinases Cleave Bacterial Flagellin, Abrogating Its Host Response-Inducing Activity. *J. Immunol.* **2004**, *172*, 509–515. [CrossRef]
6. Wiedow, O.; Meyer-Hoffert, U. Neutrophil Serine Proteases: Potential Key Regulators of Cell Signalling during Inflammation. *J. Intern. Med.* **2005**, *257*, 319–328. [CrossRef]
7. Bellani, G.; Laffey, J.G.; Pham, T.; Fan, E.; Brochard, L.; Esteban, A.; Gattinoni, L.; Van Haren, F.; Larsson, A.; McAuley, D.F.; et al. Epidemiology, Patterns of Care, and Mortality for Patients with Acute Respiratory Distress Syndrome in Intensive Care Units in 50 Countries. *JAMA* **2016**, *315*, 788–800. [CrossRef]
8. Walkey Allan, J.; Summer, R.; Ho, V.; Alkana, P. Acute Respiratory Distress Syndrome: Epidemiology and Management Approaches. *Clin. Epidemiol.* **2012**, *4*, 159. [CrossRef]
9. Faurschou, M.; Borregaard, N. Neutrophil Granules and Secretory Vesicles in Inflammation. *Microbes Infect.* **2003**, *5*, 1317–1327. [CrossRef]
10. Stapels, D.A.; Geisbrecht, B.V.; Rooijakkers, S.H. Neutrophil Serine Proteases in Antibacterial Defense. *Curr. Opin. Microbiol.* **2015**, *23*, 42–48. [CrossRef]
11. Korkmaz, B.; Horwitz, M.S.; Jenne, D.E.; Gauthier, F. Neutrophil Elastase, Proteinase 3, and Cathepsin G as Therapeutic Targets in Human Diseases. *Pharmacol. Rev.* **2010**, *62*, 726–759. [CrossRef]
12. Eyles, J.L.; Roberts, A.W.; Metcalf, D.; Wicks, I.P. Granulocyte Colony-Stimulating Factor and Neutrophils—Forgotten Mediators of Inflammatory Disease. *Nat. Clin. Pract. Rheumatol.* **2006**, *2*, 500. [CrossRef]

13. Kettritz, R. Neutral Serine Proteases of Neutrophils. *Immunol. Rev.* **2016**, *273*, 232–248. [CrossRef]
14. Turner, A.; Karube, I.; Wilson, G.S. *Biosensors: Fundamentals and Applications*; Oxford University Press: Oxford, UK, 1987.
15. Campanelli, D.; Melchior, M.; Fu, Y.; Nakata, M.; Shuman, H.; Nathan, C.; Gabay, J.E. Cloning of Cdna for Proteinase 3: A Serine Protease, Antibiotic, and Autoantigen from Human Neutrophils. *J. Exp. Med.* **1990**, *172*, 1709–1715. [CrossRef]
16. Perera, N.C.; Schilling, O.; Kittel, H.; Back, W.; Kremmer, E.; Jenne, D.E. Nsp4, an Elastase-Related Protease in Human Neutrophils with Arginine Specificity. *Proc. Natl. Acad. Sci. USA* **2012**, *109*, 6229–6234. [CrossRef]
17. Campbell, E.J.; Campbell, M.A.; Owen, C.A. Bioactive Proteinase 3 on the Cell Surface of Human Neutrophils: Quantification, Catalytic Activity, and Susceptibility to Inhibition. *J. Immunol.* **2000**, *165*, 3366–3374. [CrossRef]
18. Boxio, R.; Wartelle, J.; Nawrocki-Raby, B.; Lagrange, B.; Malleret, L.; Hirche, T.; Taggart, C.; Pacheco, Y.; Devouassoux, G.; Bentaher, A. Neutrophil Elastase Cleaves Epithelial Cadherin in Acutely Injured Lung Epithelium. *Respir. Res.* **2016**, *17*, 129. [CrossRef]
19. Griffith, M.E.; Coulthart, A.; Pemberton, S.; George, A.J.; Pusey, C.D. Anti-Neutrophil Cytoplasmic Antibodies (Anca) from Patients with Systemic Vasculitis Recognize Restricted Epitopes of Proteinase 3 Involving the Catalytic Site. *Clin. Exp. Immunol.* **2001**, *123*, 170–177. [CrossRef]
20. Corbel, M.; Boichot, E.; Lagente, V. Role of Gelatinases Mmp-2 and Mmp-9 in Tissue Remodeling Following Acute Lung Injury. *Braz. J. Med. Biol. Res.* **2000**, *33*, 749–754. [CrossRef]
21. Xu, F.; Hu, Y.; Zhou, J.; Wang, X. Mesenchymal Stem Cells in Acute Lung Injury: Are They Ready for Translational Medicine? *J. Cell. Mol. Med.* **2013**, *17*, 927–935. [CrossRef]
22. Chu, K.-E.; Fong, Y.; Wang, D.; Chen, C.F.; Yeh, D.Y.-W. Pretreatment of a Matrix Metalloproteases Inhibitor and Aprotinin Attenuated the Development of Acute Pancreatitis-Induced Lung Injury in Rat Model. *Immunobiology* **2018**, *223*, 64–72. [CrossRef] [PubMed]
23. Greenlee, K.J.; Werb, Z.; Kheradmand, F. Matrix Metalloproteinases in Lung: Multiple, Multifarious, and Multifaceted. *Physiol. Rev.* **2007**, *87*, 69–98. [CrossRef]
24. Pugin, J.; Widmer, M.C.; Kossodo, S.; Liang, C.M.; Preas, H.; Suffredini, A.F. Human Neutrophils Secrete Gelatinase B in Vitro and in Vivo in Response to Endotoxin and Proinflammatory Mediators. *Am. J. Respir. Cell Mol. Biol.* **1999**, *20*, 458–464. [CrossRef]
25. Wang, D.; Yuan, Y.; Zheng, Y.; Chai, Y.; Yuan, R. An Electrochemical Peptide Cleavage-Based Biosensor for Matrix Metalloproteinase-2 Detection with Exonuclease Iii-Assisted Cycling Signal Amplification. *Chem. Commun.* **2016**, *52*, 5943–5945. [CrossRef]
26. Sheikhzadeh, E.; Beni, V.; Zourob, M. Nanomaterial application in bio/sensors for the detection of infectious diseases. *Talanta* **2021**, *230*, 122026. [CrossRef]
27. Wignarajah, S.; Suaifan, G.A.; Bizzarro, S.; Bikker, F.J.; Kaman, W.E.; Zourob, M. Colorimetric Assay for the Detection of Typical Biomarkers for Periodontitis Using a Magnetic Nanoparticle Biosensor. *Anal. Chem.* **2015**, *87*, 12161–12168. [CrossRef] [PubMed]
28. Wanger, J.; Clausen, J.L.; Coates, A.; Pedersen, O.F.; Brusasco, V.; Burgos, F.; Casaburi, R.; Crapo, R.; Enright, P.; Van Der Grinten, C.P.; et al. Standardisation of the Measurement of Lung Volumes. *Eur. Respir. J.* **2005**, *26*, 511–522. [CrossRef]
29. Meade, M.O.; Cook, D.J.; Guyatt, G.H.; Slutsky, A.S.; Arabi, Y.M.; Cooper, D.J.; Davies, A.R.; Hand, L.E.; Zhou, Q.; Thabane, L.; et al. Ventilation Strategy Using Low Tidal Volumes, Recruitment Maneuvers, and High Positive End-Expiratory Pressure for Acute Lung Injury and Acute Respiratory Distress Syndrome: A Randomized Controlled Trial. *JAMA* **2008**, *299*, 637–645. [CrossRef]
30. Slutsky, A.S. Lung Injury Caused by Mechanical Ventilation. *Chest* **1999**, *116*, S9. [CrossRef]
31. Gajic, O.; Dara, S.I.; Mendez, J.L.; Adesanya, A.O.; Festic, E.; Caples, S.M.; Rana, R.; StSauver, J.L.; Lymp, J.F.; Afessa, B.; et al. Ventilator-Associated Lung Injury in Patients without Acute Lung Injury at the Onset of Mechanical Ventilation. *Crit. Care Med.-Baltim.* **2004**, *32*, 1817–1824. [CrossRef]
32. Younan, D.; Griffin, R.; Zaky, A.; Pittet, J.F.; Camins, B. Burn Patients with Infection-Related Ventilator Associated Complications Have Worse Outcomes Compared to Those without Ventilator Associated Events. *Am. J. Surg.* **2018**, *215*, 678–681. [CrossRef]
33. Herasevich, V.; Yilmaz, M.; Khan, H.; Hubmayr, R.D.; Gajic, O. Validation of an Electronic Surveillance System for Acute Lung Injury. *Intensive Care Med.* **2009**, *35*, 1018–1023. [CrossRef]
34. Sink, T.D.; Lochmann, R.T.; Fecteau, K.A. Validation, Use, and Disadvantages of Enzyme-Linked Immunosorbent Assay Kits for Detection of Cortisol in Channel Catfish, Largemouth Bass, Red Pacu, and Golden Shiners. *Fish Physiol. Biochem.* **2008**, *34*, 95–101. [CrossRef]
35. Gasanov, U.; Hughes, D.; Hansbro, P.M. Methods for the Isolation and Identification of Listeria Spp. And Listeria Monocytogenes: A Review. *FEMS Microbiol. Rev.* **2005**, *29*, 851–875. [CrossRef]
36. Eissa, S.; Zourob, M. Ultrasensitive Peptide-Based Multiplexed Electrochemical Biosensor for the Simultaneous Detection of Listeria Monocytogenes and Staphylococcus Aureus. *Microchim. Acta* **2020**, *187*, 486. [CrossRef] [PubMed]
37. Eissa, S.; Zourob, M. A Dual Electrochemical/Colorimetric Magnetic Nanoparticle/Peptide-Based Platform for the Detection of Staphylococcus Aureus. *Analyst* **2020**, *145*, 4606–4614. [CrossRef] [PubMed]
38. Li, Y.; Schluesener, H.J.; Xu, S. Gold Nanoparticle-Based Biosensors. *Gold Bull.* **2010**, *43*, 29–41. [CrossRef]
39. Driscoll, K.E. Tnfα and Mip-2: Role in Particle-Induced Inflammation and Regulation by Oxidative Stress. *Toxicol. Lett.* **2000**, *112*, 177–183. [CrossRef] [PubMed]

40. Jiang, D.; Liang, J.; Fan, J.; Yu, S.; Chen, S.; Luo, Y.; Prestwich, G.D.; Mascarenhas, M.M.; Garg, H.G.; Quinn, D.A.; et al. Regulation of Lung Injury and Repair by Toll-Like Receptors and Hyaluronan. *Nat. Med.* **2005**, *11*, 1173. [CrossRef]
41. Massey, V.L.; Poole, L.G.; Siow, D.L.; Torres, E.; Warner, N.L.; Schmidt, R.H.; Ritzenthaler, J.D.; Roman, J.; Arteel, G.E. Chronic Alcohol Exposure Enhances Lipopolysaccharide-Induced Lung Injury in Mice: Potential Role of Systemic Tumor Necrosis Factor-Alpha. *Alcohol. Clin. Exp. Res.* **2015**, *39*, 1978–1988. [CrossRef]
42. Bohr, A.; Tsapis, N.; Andreana, I.; Chamarat, A.; Foged, C.; Delomenie, C.; Noiray, M.; El Brahmi, N.; Majoral, J.P.; Mignani, S.; et al. Anti-Inflammatory Effect of Anti-Tnf-A Sirna Cationic Phosphorus Dendrimer Nanocomplexes Administered Intranasally in a Murine Acute Lung Injury Model. *Biomacromolecules* **2017**, *18*, 2379–2388. [CrossRef]
43. Xu, Y.; Ito, T.; Fushimi, S.; Takahashi, S.; Itakura, J.; Kimura, R.; Sato, M.; Mino, M.; Yoshimura, A.; Matsukawa, A. Spred-2 Deficiency Exacerbates Lipopolysaccharide-Induced Acute Lung Inflammation in Mice. *PLoS One* **2014**, *9*, e108914. [CrossRef]
44. Perl, M.; Lomas-Neira, J.; Chung, C.S.; Ayala, A. Epithelial Cell Apoptosis and Neutrophil Recruitment in Acute Lung Injury—a Unifying Hypothesis? What We Have Learned from Small Interfering Rnas. *Mol. Med.* **2008**, *14*, 465–475. [CrossRef] [PubMed]
45. Craven, T.H.; Avlonitis, N.; McDonald, N.; Walton, T.; Scholefield, E.; Akram, A.R.; Walsh, T.S.; Haslett, C.; Bradley, M.; Dhaliwal, K. Super-Silent Fret Sensor Enables Live Cell Imaging and Flow Cytometric Stratification of Intracellular Serine Protease Activity in Neutrophils. *Sci. Rep.* **2018**, *8*, 13490. [CrossRef] [PubMed]
46. Popow-Stellmaszyk, J.; Bajorowicz, B.; Malankowska, A.; Wysocka, M.; Klimczuk, T.; Zaleska-Medynska, A.; Lesner, A. Design, Synthesis, and Enzymatic Evaluation of Novel Zno Quantum Dot-Based Assay for Detection of Proteinase 3 Activity. *Bioconjugate Chem.* **2018**, *29*, 1576–1583. [CrossRef] [PubMed]
47. Jing, P.; Yi, H.; Xue, S.; Yuan, R.; Xu, W. A 'Signal on-Off'electrochemical Peptide Biosensor for Matrix Metalloproteinase 2 Based on Target Induced Cleavage of a Peptide. *RSC Adv.* **2015**, *5*, 65725–65730. [CrossRef]
48. Kou, B.-B.; Zhang, L.; Xie, H.; Wang, D.; Yuan, Y.-L.; Chai, Y.-Q.; Yuan, R. DNA Enzyme-Decorated DNA Nanoladders as Enhancer for Peptide Cleavage-Based Electrochemical Biosensor. *ACS Appl. Mater. Interfaces* **2016**, *8*, 22869–22874. [CrossRef]
49. Wang, H.; Ma, Z. A Novel Strategy for Improving Amperometric Biosensor Sensitivity Using Dual-Signal Synergistic Effect for Ultrasensitive Detection of Matrix Metalloproteinase-2. *Sens. Actuators B Chem.* **2018**, *266*, 46–51. [CrossRef]
50. Kou, B.-B.; Chai, Y.-Q.; Yuan, Y.-L.; Yuan, R. Ptnps as Scaffolds to Regulate Interenzyme Distance for Construction of Efficient Enzyme Cascade Amplification for Ultrasensitive Electrochemical Detection of Mmp-2. *Anal. Chem.* **2017**, *89*, 9383–9387. [CrossRef]
51. Wang, H.; Ma, Z.; Han, H. A Novel Impedance Enhancer for Amperometric Biosensor Based Ultrasensitive Detection of Matrix Metalloproteinase-2. *Bioelectrochemistry* **2019**, *130*, 107324. [CrossRef]
52. Saikiran, M.; Sato, D.; Pandey, S.S.; Hayase, S.; Kato, T. Efficient near Infrared Fluorescence Detection of Elastase Enzyme Using Peptide-Bound Unsymmetrical Squaraine Dye. *Bioorganic Med. Chem. Lett.* **2017**, *27*, 4024–4029. [CrossRef] [PubMed]
53. Edwards, J.V.; Prevost, N.; French, A.; Concha, M.; DeLucca, A.; Wu, Q. Nanocellulose-Based Biosensors: Design, Preparation, and Activity of Peptide-Linked Cotton Cellulose Nanocrystals Having Fluorimetric and Colorimetric Elastase Detection Sensitivity. *Engineering* **2013**, *5*, 20. [CrossRef]
54. Folch, E.; Salas, A.; Panés, J.; Gelpí, E.; Roselló-Catafau, J.; Anderson, D.C.; Navarro, S.; Piqué, J.M.; Fernández-Cruz, L.; Closa, D. Role of P-Selectin and Icam-1 in Pancreatitis-Induced Lung Inflammation in Rats: Significance of Oxidative Stress. *Ann. Surg.* **1999**, *230*, 792. [CrossRef]
55. Kaman, W.E.; Voskamp-Visser, I.; de Jongh, D.M.; Endtz, H.P.; van Belkum, A.; Hays, J.P.; Bikker, F.J. Evaluation of a D-Amino-Acid-Containing Fluorescence Resonance Energy Transfer Peptide Library for Profiling Prokaryotic Proteases. *Anal. Biochem.* **2013**, *441*, 38–43. [CrossRef] [PubMed]
56. Fujinaga, M.; Chernaia, M.M.; Halenbeck, R.; Koths, K.; James, M.N. The Crystal Structure of Pr3, a Neutrophil Serine Proteinase Antigen of Wegener's Granulomatosis Antibodies. *J. Mol. Biol.* **1996**, *261*, 267–278. [CrossRef] [PubMed]
57. Lu, G.; Zheng, M.; Zhu, Y.; Sha, M.; Wu, Y.; Han, X. Selection of Peptide Inhibitor to Matrix Metalloproteinase-2 Using Phage Display and Its Effects on Pancreatic Cancer Cell Lines Panc-1 and Cfpac-1. *Int. J. Biol. Sci.* **2012**, *8*, 650. [CrossRef]
58. Avlonitis, N.; Debunne, M.; Aslam, T.; McDonald, N.; Haslett, C.; Dhaliwal, K.; Bradley, M. Highly Specific, Multi-Branched Fluorescent Reporters for Analysis of Human Neutrophil Elastase. *Org. Biomol. Chem.* **2013**, *11*, 4414–4418. [CrossRef]
59. Alhogail, S.; Suaifan, G.A.; Zourob, M. Rapid Colorimetric Sensing Platform for the Detection of Listeria Monocytogenes Foodborne Pathogen. *Biosens. Bioelectron.* **2016**, *86*, 1061–1066. [CrossRef]

 *biosensors* 

Article

Enhanced Competitive Immunomagnetic Beads Assay Assisted with PAMAM-Gold Nanoparticles Multi-Enzyme Probes for Detection of Deoxynivalenol

Kun Zeng [1], Jian Yang [1], Hao Su [1], Sheng Yang [1], Xinkai Gu [1], Zhen Zhang [1] and Hongjun Zhao [2,*]

[1] School of the Environment and Safety Engineering, Jiangsu University, Zhenjiang 212013, China
[2] Department of Pulmonary and Critical Care Medicine, The Quzhou Affiliated Hospital of Wenzhou Medical University, Quzhou People's Hospital, 100 Minjiang Ave, Quzhou 324000, China
* Correspondence: zhaohongjun@wmu.edu.cn; Tel.: +86-570-3121509

Abstract: Contamination of deoxynivalenol (DON) in grains has attracted widespread concern. It is urgently needed to develop a highly sensitive and robust assay for DON high-throughput screening. Antibody against DON was assembled on the surface of immunomagnetic beads orientationally by the aid of Protein G. AuNPs were obtained under the scaffolding of poly(amidoamine) dendrimer (PAMAM). DON-horseradish peroxidase (HRP) was combined on the periphery of AuNPs/PAMAM by a covalent link to develop DON-HRP/AuNPs/PAMAM. Magnetic immunoassay based on DON-HRP/AuNPs/PAMAM was optimized and that based on DON-HRP/AuNPs and DON-HRP was adopted as comparison. The limits of detection (LODs) were 0.447 ng/mL, 0.127 ng/mL and 0.035 ng/mL for magnetic immunoassays based on DON-HRP, DON-HRP/Au and DON-HRP/Au/PAMAM, respectively. Magnetic immunoassay based on DON-HRP/AuNPs/PAMAM displayed higher specificity towards DON and was utilized to analyze grain samples. The recovery for the spiked DON in grain samples was 90.8–116.2% and the method presented a good correlation with UPLC/MS. It was found that the concentration of DON was in the range of ND-3.76 ng/mL. This method allows the integration of dendrimer–inorganic NPs with signal amplification properties for applications in food safety analysis.

Keywords: deoxynivalenol; gold nanoparticles; PAMAM; multi-enzyme probes

Citation: Zeng, K.; Yang, J.; Su, H.; Yang, S.; Gu, X.; Zhang, Z.; Zhao, H. Enhanced Competitive Immunomagnetic Beads Assay Assisted with PAMAM-Gold Nanoparticles Multi-Enzyme Probes for Detection of Deoxynivalenol. *Biosensors* **2023**, *13*, 536. https://doi.org/10.3390/bios13050536

Received: 16 March 2023
Revised: 3 May 2023
Accepted: 5 May 2023
Published: 10 May 2023

1. Introduction

Mycotoxins are secondary metabolites produced by fungi, which often pollute crops in the period of growth, harvest and storage [1]. Among hundreds of mycotoxins, deoxynivalenol (DON) is isolated from corn contaminated by Fusarium graminearum and is one of the most common contaminated toxins in food crops. It was reported that 90% of the samples infected with mycotoxins contained DON [2]. DON can directly pollute crops and then influence the health of people or animals through the food chain, resulting in vomiting, anorexia, gastroenteritis, diarrhea, slow growth and even immune suppression or blood disease [3–5]. European Union (EU) establishes the maximum limit of DON of 750 μg/kg for cereal flour and 1750 μg/kg for unprocessed wheat [6], while the maximum limit of DON in cereals and cereal products is set as 1000 μg/kg in China [7]. To date, numerous analytical methods have been developed to detect DON in food, including high-performance liquid chromatography [8,9], liquid chromatography-tandem mass spectrometry (LC-MS) [10,11] and immunoassays [12]. Technical methods always rely on professional faculty, expensive instruments and tedious sample preparations, which would hinder their applications in the rapid screening of high-throughput samples.

Immunoassays have attracted concerns in the food safety field by the virtues of simple operations, low costs and high throughput. After specific binding between antibodies and antigens, various reporters, such as natural enzymes, fluorescein and quantum dots, have been utilized to output signals, where higher signal value will always be related to higher sensitivity. Among these reporters, enzymes are the most commonly used labels since their amazing catalytic activities would produce strong output values in a short time. To improve the performance of analytical methods further, numerous enzymes would be assembled together to form an enzyme complex to amplify the detection signals through an antigen–antibody system [13], biotin–avidin system [14] or polymers [15]. Moreover, emerging nanomaterials provide new strategies to develop an enzyme complex, which could reduce diffusion limitations and maximize the functional surface area to increase enzyme loading [16]. Gold nanoparticles (AuNPs) [17], carbon nanotubes (CNTs) [18,19] and so on have been explored. In our previous study, AuNPs and carbon nanotubes were adopted as carriers for loading horseradish peroxidase (HRP) [18]. The limit of detection (LOD) in the immunoassay based on CNTs multi-enzyme nanomaterials was improved up to 37-fold compared with conventional immunoassay, in which the higher catalytic efficiency of multi-enzyme nanomaterials played a crucial role. Although these nanoparticles served as excellent biocompatible surfaces for the immobilization of proteins, they had a strong tendency to agglomerate and non-specific binding, resulting in higher background and affecting their utility in analytical methods [20].

Dendrimers are a specific class of highly branched macromolecules with well-defined three-dimensional architecture [21]. The branching units radiate from the central monomer (i.e., core) to form the multifunctional macromolecules, in which the diameter of dendrimers increases linearly along with each generation, and the number of end groups increases exponentially [22]. Dendrimers would facilitate the controlled synthesis of inorganic nanoparticles as nanoreactors. Moreover, the high number of terminal groups at the periphery could protect the particles and increase their solubility and stability under variable conditions [23]. Furthermore, the functionalized end groups would make them the ideal host for various biological molecules, such as enzymes and antibodies. Given the structural versatility of dendrimers, the application of dendrimer–inorganic NPs has grown immensely in the field of catalysis [24], imaging [25], biomedicine [26] and environment [27]. Nevertheless, it is still less studied to apply dendrimer–inorganic NPs in immunoassay, especially for signal amplification strategies.

In this study, a competitive immunomagnetic beads assay with poly (amidoamine) dendrimer (PAMAM)—gold nanoparticles multi-enzyme probes was developed to determine the concentration of DON in food. Magnetic beads were decorated with Protein G to immobilize the antibody against DON orientationally. DON-HRP and AuNPs/PAMAM were synthesized and DON-HRP was assembled on the periphery of AuNPs/PAMAM by a covalent link. The novel magnetic immunoassay based on AuNPs/PAMAM multi-enzyme probes was established and optimized. The grain samples in local markets were collected and detected by the new method.

2. Materials and Methods

2.1. Reagents and Materials

Protein G, HAuCL4, Ovalbumin (OVA), Bovine Serum Albumin (BSA), 3,3′,5,5′-Tetramethylbenzidine (TMB), HRP, 1,1′-carbonyl-diimidazole (CDI), 1-(3-(Dimethylamino)-propyl)-3-ethylcarbodiimide hydrochloride (EDC) and N-hydroxysulfosuccinimide (NHS) were purchased from Sigma-Aldrich (St. Louis, MO, USA). The standards of DON, 3-acetyldeoxynivalenol (3-AcDON), 15-acetyldeoxynivalenol (15-AcDON), T-2 toxin, zearalenone (ZEN), ochratoxin (OTA), aflatoxin B1 (AFB1) and aflatoxin M1 (AFM1) were obtained from Beijing Zhongke Quality Inspection Biotechnology Co., Ltd., (Beijing, China). PAMAM dendrimer (ethylenediamine core, generation 4.0, amino-terminated, 5 wt% in methanol) was purchased from Weihai Chenyuan Molecular New Materials Co., Ltd. (Weihai, China). Monoclonal antibody against DON was obtained from Shenzhen Kejie Industrial Development Co., Ltd., (Shenzhen, China). Other reagents were bought from Sinopharm Chemical Reagent Beijing Co., Ltd. (Beijing, China). Immunomagnetic beads (IMBs) were purchased from Suzhou Weidu Biological Co., Ltd. (Suzhou, China).

2.2. Preparation of DON-HRP Colorimetric Probe

DON-HRP was developed according to the previous experiment with minor modifications [28]. About 10 mg of CDI was dissolved in 1 mL of anhydrous tetrahydrofuran (THF) and then 2 mg of DON was added. After stirring for 2 h at 25 °C, the organic solvent was removed through a vacuum pump. The products were collected and leached out in 200 μL DMSO. The mixture was added drop by drop into 1 mL of PBS containing 3 mg of HRP. After a full reaction at 4 °C for 20 h, the solution was dialyzed against 0.01 mol/L PBS at 4 °C for 48 h and the buffer was changed every 4 h. The products, DON-HRP, were stored at 4 °C until further use within two weeks.

2.3. Preparation of Colorimetric Probe Based on AuNPs

AuNPs with an average size of 20 nm were prepared according to the previous experiment [17]. AuNPs solution was adjusted to pH 8.0 with 0.1 mol/L K_2CO_3. Then 100 μL DON-HRP (1.5 mg/mL) and 1.0 mL AuNPs solution were reacted at 25 °C for 0.5 h with stirring slowly. Next, to occupy the non-specific binding site on AuNPs, 10 μL BSA (20%, m/v) was mixed with the above solution for 30 min. Finally, DON-HRP/AuNPs were gathered after centrifugation at 11,000× g for 15 min and then redissolved in 1 mL of PB buffer (0.01 mol/L, containing 1% PEG20000), stored at 4 °C and used within two weeks.

2.4. Preparation of Colorimetric Probe Based on AuNPs/PAMAM

First, 1 mL of 30 mM HAuCL4 solution and 1.5 mL of PAMAM (1.0 wt%) were mixed for 60 min with vigorous stirring. Then 1 mL of 0.5 mol/L NaBH4 solution was added into the solution and stirred for another 30 min. The mixture was centrifuged at 12,000× g for 10 min to remove excess PAMAM and NaBH4 and then the precipitate, AuNPs/PAMAM, was resuspended in 1 mL ddH_2O. AuNPs/PAMAM was mixed with 10 mg EDC and 15 mg NHS by stirring for 60 min at 4 °C and then was centrifuged at 12,000× g for 10 min to remove the excess EDC and NHS. Furthermore, DON-HRP was added for 12 h at 4 °C and the final product, DON-HRP/AuNPs/PAMAM, was collected by centrifugation and stored at 4 °C and then resuspended in 1 mL PBS buffer containing 1% BSA. The procedure was presented in Figure 1A.

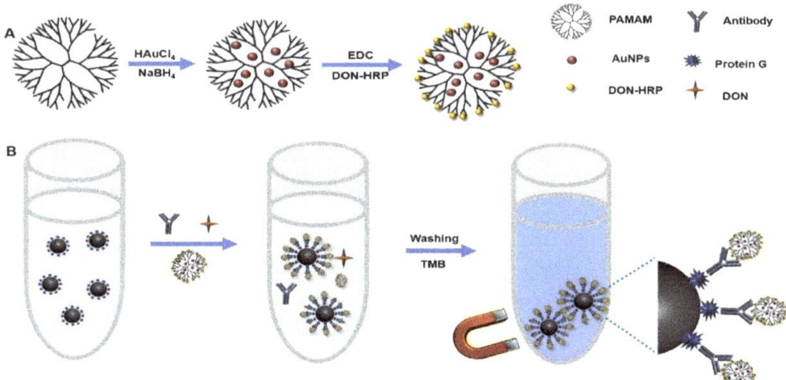

Figure 1. The diagram of magnetic immunoassay based on DON-HRP/Au/PAMAM: (**A**) the synthesis of DON-HRP/Au/PAMAM and (**B**) the flowchart of magnetic immunoassay based on DON-HRP/Au/PAMAM.

2.5. Determination of Catalytic Property of Immobilized HRP

The activities of immobilized HRP in DON-HRP, DON-HRP/AuNPs and DON-HRP/AuNPs/PAMAM were explored and compared with free HRP. After protein quantification, the final HRP concentration of 0.5 ng/mL for each multi-enzyme probe was adopted to react with TMB/H_2O_2 mixture in a 96-well microplate. Absorbances at 652 nm were recorded every 30 s at time intervals of 0~600 s. The results were analyzed and fitted by Origin software.

2.6. Immobilization of Protein G on IMBs

To prepare the active IMBs, 20 mg EDC and 35 mg NHS were dissolved in 2 mL MES buffer (50 mmol/L, pH 6.0) and then 200 μL IMBs were added with continuous stirring for 4 h at 25 °C. After being collected by a magnetic separator, the active IMBs were reacted with Protein G overnight at 4 °C under gentle shaking. The products, IMBs-Protein G, were washed with MES buffer for three times and then resuspended in 1 mL PBS containing 2% skim milk and stored at 4 °C.

2.7. One-Step Magnetic Immunoassay with Different Signal Probes

In one plastic tube, the following reagents were added in sequence: 50 μL of IMBs-Protein G, 50 μL of mAb against DON, 50 μL of DON standards or samples and 50 μL of diluted probe (DON-HRP, DON-HRP/AuNPs and DON-HRP/AuNPs/PAMAM). After mixed fully and reacted at 37 °C for 30 min, the mixture was separated by a magnet and washed three times with 500 μL PBST. To develop the colorimetric reaction, freshly prepared TMB solution was added in drops into each tube for 20 min at 37 °C (Figure 1B). After stopping the reaction by H_2SO_4 (2 mol/L), the absorbance of the supernatant was recorded by a microplate reader at 450 nm.

2.8. Optimization of Magnetic Immunoassays with Multi-Enzyme Probes

The usage of the key elements in magnetic immunoassays was investigated. Different usage of IMBs (20, 30 and 40 μL), dilution of antibody (1:500, 1:1000 and 1:200) and dilution of multi-enzyme probes (1:50, 1:100 and 1:150) were matched with each other, and the absorbances at 0 ng/mL and 5 ng/mL of DON were recorded.

To optimize the reaction condition, solutions with different factors were utilized to prepare the analyte standards. For pH, the values were 6.5, 7.0, 7.4, 8.0 and 9.0. The concentrations of Na^+ were 0, 0.01, 0.05, 0.1 and 0.2 mol/L, and acetonitrile (m/v) was 0, 5%, 10%, 20% and 40%. The maximum absorbances (B0) were recorded, and IC50 was the concentration at which 50% of the antibodies were bound to the analyte.

2.9. Samples Preparation and Analysis

Samples of rice, corn and wheat from local markets were crushed with a pulverizer and passed a 20-mesh screen. Then, 5 g of the sample powers in a 50 mL polypropylene tube was mixed with 20 mL of PBS containing 10% acetonitrile. After vigorous mixing for 10 min, the extracts were utilized for analysis by the developed method. A total of 30 natural samples were tested and compared by the new method and ultra-high-performance liquid chromatography-tandem mass spectrometry (UPLC-MS/MS).

For UPLC-MS/MS detection, we followed the procedures in previous reports [29]. First, 5 g of the fine grain sample in a 50 mL polypropylene tube was mixed with 25 mL of acetonitrile/water (84:16, v/v). The sample was extracted on a shaker for 30 min and centrifuged for 10 min at 5000× g. Then 2 mL of the supernatant was passed through a solid phase extraction column, the filtrate was dried by N2 and the product was dissolved in 1 mL of acetonitrile/water (1:9, v/v). The solution was injected into Agilent 1290 UPLC coupled with ABsciex 4500 MS. The gradient elution program was carried out. Each sample was evaluated three times in duplicate.

3. Results

3.1. Construction, Optimization and Characterization of Multi-Enzyme Probes

G4 PAMAM with 64 NH2 terminals was utilized as the scaffold to guest AuNPs in situ growth in this study (Figure 1A). The optimal usage of PAMAM and HAuCL4 was explored. It was observed that the absorbance changed with the volume ratio of HAuCL4 and PAMAM (1:0.5, 1:1 and 1:1.5) when 50 µg of DON-HRP was as used. The OD value at the ratio of 1:0.5 was lower than that at the ratio of 1:1 and 1:1.5 (Figure 2a). According to TEM images (Figure 2b), AuNPs/PAMAM showed an average size of 5~8 nm, smaller than conventional AuNPs (20 nm, Figure S1), and presented a dendrimer coating around metal nanoparticles. AuNPs/PAMAM at the ratio of 1:1 processed better mono-dispersity and excess PAMAM may result in the clustering of nanoparticles. Therefore, the volume ratio of HAuCL4 and PAMAM was fixed at 1:1 in the next experiment. With the help of the amino group in the periphery of PAMAM, enzyme molecules could be conjugated by covalent crosslinking to form the multi-enzymes complex. More enzymes loading on the nanoparticles carrier would contribute to the improvement of signals, which was confirmed in the results (Figure 2c). With the increasing HRP content, the absorbance improved significantly and a platform was observed over 40 µg of DON-HRP. Hence, 40 µg of DON-HRP was chosen in the next experiment.

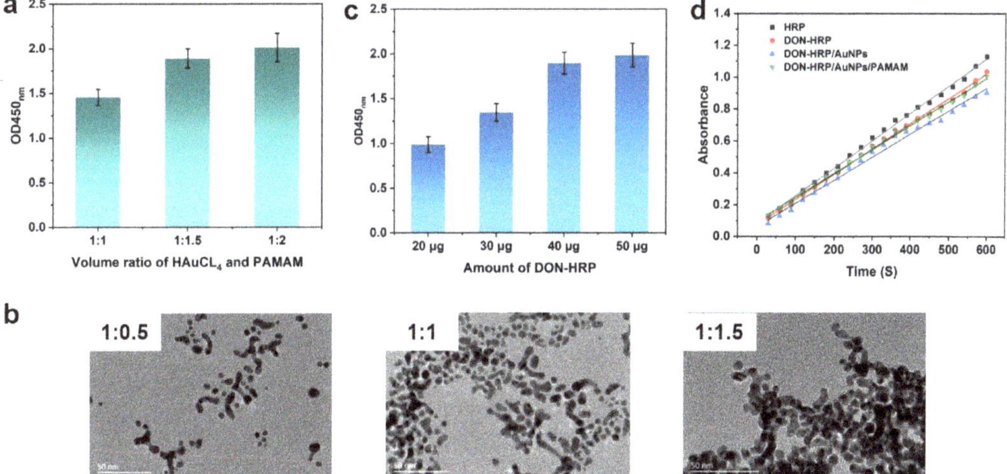

Figure 2. Development of multi-enzymes particles based on Au/PAMAM: (**a**) absorbance at different volume ratios of HAuCl4 and PAMAM; (**b**) TEM images of Au/PAMAM at different volume ratios; (**c**) optimization of DON-HRP; (**d**) catalytic activities for DON-HRP, DON-HRP/AuNPs and DON-HRP/AuNPs/PAMAM.

Furthermore, the catalytic property of multi-enzyme particles was measured using TMB/H_2O_2 as a substrate. The concentration of HRP in multi-enzyme probes was determined through the BCA kit, and 1 µg of absolute content HRP in each multi-enzyme probe was utilized to evaluate the catalytic property. In comparison with the initial slope of catalysis kinetics, the activities of a single immobilized HRP relative to free HRP could be calculated (Figure 2d). Then, the relative activities of immobilized HRP were 91.86%, 83.72% and 88.63% for DON-HRP, DON-HRP/AuNPs and DON-HRP/AuNPs/PAMAM, respectively.

3.2. *Optimization of IMBs Modified with Protein G*

Protein G could help antigen-binding (Fab) regions of antibody orientation on the surface as far as possible (Figure 1B), which would confirm the activity of antibodies. To obtain the optimal condition, the usage of Protein G was tested when 1:1000 dilution of DON-HRP and 50 µL of IMBs were utilized. As shown in Figure 3a, the OD value increased with the concentration of Protein G below 50 µg per test, which mean the maximum loading for Protein G. To block the non-specific combination on IMBs, some reagents, including

BSA, gelatin and skim milk powder, were added into the solution to reduce the background, in which 2% skim milk displayed the best performance (Figure 3b).

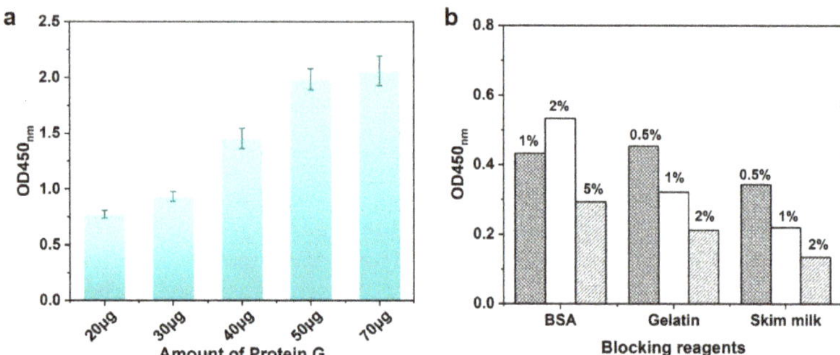

Figure 3. The optimization of IMB-Protein G: (**a**) optimization of usage of Protein G and (**b**) optimization of blocking reagents.

3.3. Optimization of Magnetic Immunoassay Based on Multi-Enzyme Probes

Bottomed on the reaction elements obtained above, magnetic immunoassays were established (Figure 1B). Firstly, the usage of these elements, including IMB-Protein G, an antibody against DON and multi-enzyme probes, should be explored for the best performance. Because of the competitive format, the absorbances at 0 ng/mL and 5 ng/mL of DON were recorded and the inhibition rates at 5 ng/mL of DON were calculated, in which higher absorbance at 0 ng/mL of DON and greater inhibition rates would signify the better sensitivity. According to this criterion, the optimal conditions were 40 μL of IMB-Protein G, 1:1000 dilution of antibody and 1:100 dilution of DON-HRP/AuNPs for magnetic immunoassay based on DON-HRP/AuNPs (Table S1), while 40 μL of IMB-Protein G, 1:2000 dilution of antibody and 1:50 dilution of Au/DON-HRP/PAMAM for another magnetic immunoassay (Table S2).

Furthermore, some physicochemical factors should be evaluated. Two parameters, B0 and B0/IC50, were used to estimate the performance of these methods, in which a higher value of B0/IC50 would be associated with higher sensitivity. As shown in Figure 4a,d, it was observed that the value of B0 exhibited higher absorbance when pH exceeded 7.0 in both magnetic immunoassays. The optimal condition was at pH 7.0 and 7.4 for the two methods, respectively. The salt ion strength could interfere with the combination between antibodies and antigens but also would induce the gathering of some nanoparticles and then interfere with the characteristics of the analysis method. It was observed that the attendance of Na+ leads to the significant decreasing trend of the value of B0/IC50 in magnetic immunoassay based on DON-HRP/AuNPs, which means the loss of sensitivity. For magnetic immunoassay based on DON-HRP/AuNPs/PAMAM, the value of B0/IC50 presented minor fluctuations below 0.1 mol/L NaCl and decreased at 0.2 mol/L NaCl. Besides pH and ion strength, organic solvent needed to be evaluated since some organic solvents, such as acetonitrile, would be utilized to extract DON from food samples. The sensitivity of magnetic immunoassay based on DON-HRP/AuNPs would be influenced over 5% acetonitrile, while the higher tolerance of acetonitrile over 20% was recorded in that based on DON-HRP/AuNPs/PAMAM.

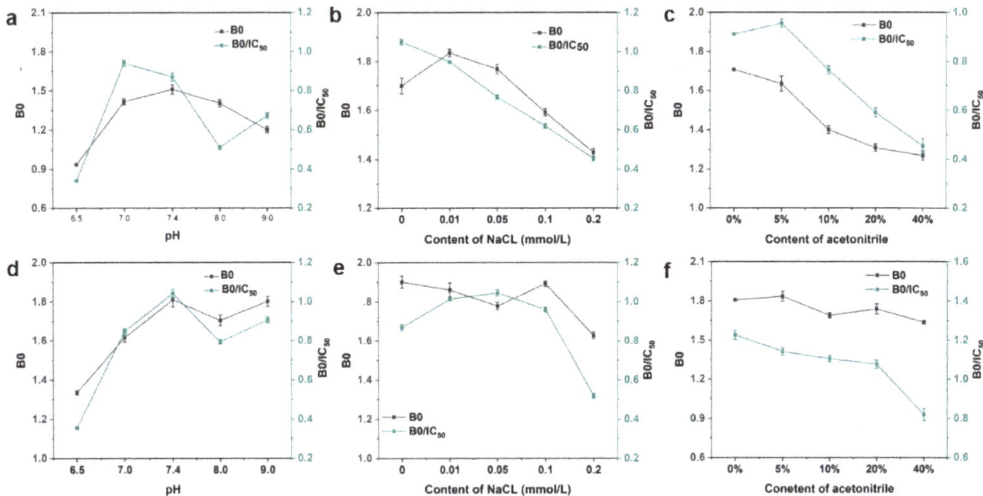

Figure 4. Optimal conditions in magnetic immunoassay based on DON-HRP/Au (**a–c**) and DON-HRP/Au/PAMAM (**d–f**): (**a**,**d**) optimization of pH, (**b**,**e**) optimization of ion strength and (**c**,**f**) optimization of organic solution.

3.4. Analytical Performance of Magnetic Immunoassays Based on Multi-Enzyme Probes

Based on the optimal condition, magnetic immunoassays were developed with the aid of DON-HRP/AuNPs and DON-HRP/AuNPs/PAMAM, and conventional magnetic immunoassay based on DON-HRP was established as comparison, and the standard curves were shown in Figure 5. It was obvious that the two magnetic immunoassays by the aid of nanoparticles shifted to the left compared with that of conventional magnetic immunoassays. After calculation, the LODs (90% inhibition concentration) were 0.447 ng/mL, 0.127 ng/mL and 0.035 ng/mL for magnetic immunoassays based on DON-HRP, DON-HRP/Au and DON-HRP/Au/PAMAM, respectively. The linear ranges (20–80% inhibition concentration) were 1.72–12.11 ng/mL, 0.932–5.68 ng/mL and 0.13–3.12 ng/mL for the three methods. In comparison with conventional magnetic immunoassays, that based on DON-HRP/AuNPs/PAMAM showed the highest sensitivity with a 13-fold improvement of LOD, while the LOD of the method based on DON-HRP/Au was upgraded 4-fold.

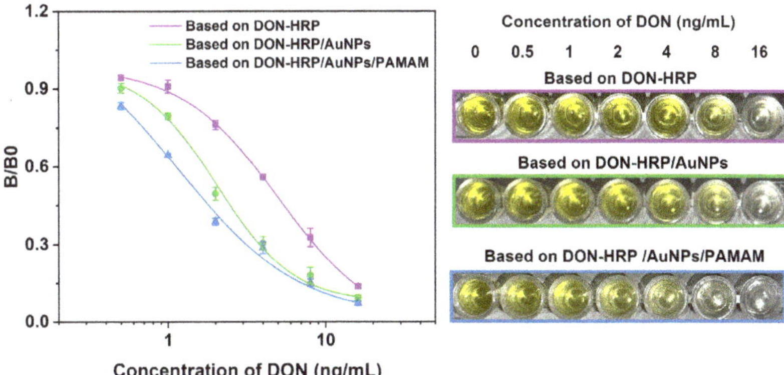

Figure 5. Standard curves of magnetic immunoassays based on DON-HRP, DON-HRP/AuNPs and DON-HRP/AuNPs/PAMAM.

3.5. Analysis of Grain Samples

Magnetic immunoassay based on DON-HRP/AuNPs/PAMAM was utilized to determine the concentration of DON in grain samples. The selectivity of the new assay should be evaluated because of the widespread contamination of mycotoxins. The results showed that there was lower cross-reactivity with 3-Ac-DON (11.2%) and 15-Ac-DON (3.4%) and no cross-reactivity with T-2, ZEN, OTA, AFB1 and AFM1 (Figure 6). To determine the usability of the new method, rice, corn and wheat samples were spiked with various contents of DON and then tested by magnetic immunoassay based on DON-HRP/AuNPs/PAMAM. The recovery rates in rice, corn and wheat samples were 96.4–116.2%, 95.8–109.2% and 90.8–114.5%, respectively, and the coefficient variations (CVs) were in the range of 7.8–11.9% (Table 1). The higher accuracy and precision in the newly established method would guarantee reliable application in real sample analysis.

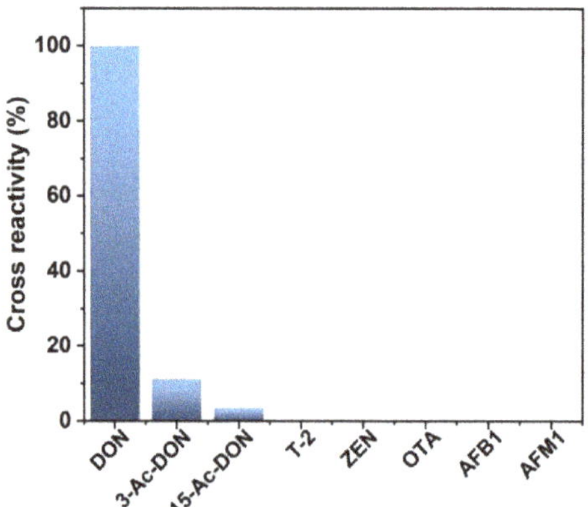

Figure 6. Cross-reactivity of magnetic immunoassays based on DON-HRP/AuNPs/PAMAM.

Table 1. Recovery and precision of the established method ($n = 4$).

	Spiked Concentration (ng/mL)	Found Concentration (ng/mL)	Recovery Rate	CV (%)
Rice	0	<LOD	-	-
	1	1.16 ± 0.09	116.2%	7.8%
	5	4.82 ± 0.53	96.4%	10.9%
	10	11.21 ± 1.29	112.1%	11.5%
Corn	0	<LOD	-	-
	1	1.09 ± 0.13	109.2%	11.9%
	5	5.43 ± 0.63	108.6%	11.6%
	10	9.58 ± 1.01	95.8%	10.5%
Wheat	0	<LOD	-	-
	1	1.06 ± 0.11	106.2%	10.4%
	5	4.54 ± 0.41	90.8%	9.0%
	10	11.45 ± 1.25	114.5%	10.9%

Furthermore, 30 grain samples spiked with a certain concentration of DON were determined by magnetic immunoassay based on DON-HRP/AuNPs/PAMAM and UPLC-MS/MS simultaneously. The results of the two methods presented a positive correlation (R = 0.9095) (Figure S2). Thirty samples (rice, corn and wheat) from local markets were analyzed (Table 2). The highest concentration of DON detected was 24.86 ng/mL in rice samples, 34.76 ng/mL in corn samples and 3.43 ng/mL in wheat samples, which were all below the MRL set by the EU and China. The detection rates were 40% in rice, 50% in corn and 40% in wheat samples.

Table 2. Measurement of DON in grain samples.

Sample Number	Results of Rice Samples (ng/mL)		Results of Corn Samples (ng/mL)		Results of Wheat Samples (ng/mL)	
	By the Established Method	By UPLC-MS/MS	By the Established Method	By UPLC-MS/MS	By the Established Method	By UPLC-MS/MS
1	ND [1]	ND	8.87	4.65	2.12	1.78
2	4.56	3.43	ND	ND	ND	ND
3	ND	ND	6.86	ND	ND	ND
4	12.98	17.87	15.76	13.34	ND	ND
5	24.86	29.86	14.76	10.89	3.43	4.65
6	ND	ND	ND	ND	ND	ND
7	ND	ND	ND	ND	3.12	4.76
8	ND	ND	34.76	35.67	ND	ND
9	ND	ND	ND	ND	6.87	5.24
10	3.24	1.68	ND	1.43	ND	ND

[1] not detected.

4. Discussion

In magnetic beads assay, antibody-decorated IMBs could not only combine with antigens or analytes as solid support but also help to extract and concentrate the analytes from the sample matrix. In conventional immobilized methods, antibodies would be linked with the active group on the surface of IMBs by covalent binding or absorbance by physical reaction through hydrophobic interaction and/or electrostatic force, where their biological activity would be partially or completely lost because of the steric hindrance and the change of the conformation of the active site during fixation [30]. Antibodies with orderly arrangement would contribute to the better performance of immunoassays. Through precise design, non-functional sites in an antibody would be adopted to attach to the surface, while recognition sites could be oriented towards the solution to ensure the sufficient exposure of functional groups. Numerous strategies had been adopted to improve antibody binding, including surface modification [31], covalent attachment [32], biotin–streptavidin interaction, DNA-directed immobilization [33] and Protein G. In this study, Protein G, derived from bacteria, was introduced as a bridge to promote the orderly arrangement of antibodies since Protein G has the ability to bind the Fc fragment of antibodies specifically [34,35].

Signal amplification strategies based on nanoparticles have attracted more attention in analysis methods. Due to the high biocompatibility and large specific surface area, AuNPs serve as excellent carriers for the immobilization of proteins. However, AuNPs always tend to aggregate while the protein binding on the surface of AuNPs will be easy to form steric hindrance, which could interfere with the combination of more enzyme molecules. Dendrimers have emerged as optimal surface capping agents given their unique chemistry and structure for the controlled synthesis of metal nanoparticles [22]. According to the structure, dendrimers with a smaller generation (i.e., 0.5–2.5 generation) are usually more flexible, while higher-generation dendrimers (i.e., 3.0–5.0 generation) tend to exhibit a denser and more globular shape. It was observed that higher-generation dendrimers have been extensively applied for in situ growth of metal NPs, due to the void spaces in their architecture [36]. Furthermore, hydroxyl-, aldehyde- and amino-terminated PAMAM presented diverse capacities to host metal nanoparticles, while amino-terminated PAMAM showed higher interaction with gold nanoclusters in comparison with hydroxyl- and aldehyde-terminated PAMAM [37]. Therefore, G4 PAMAM with NH_2 terminals was utilized as the scaffold for guest AuNPs in this study.

With the help of the amino group, more enzymes or other signal molecules could be linked with the nanostructure. Hu et al. found that electrochemiluminescence (ECL) based on PAMAM-Au-CdSe quantum dot bioprobes displayed satisfied sensitivity with the LOD of 3.7×10^{-6} ng/mL for the detection of brombuterol [38]. Moreover, it was found that AuNPs/PAMAM could be used as drugs or gene carriers by attaching various polymers or antibodies, which implicated that the higher molecule activity would be reserved by this nanostructure [25,26]. Similarly, the higher catalytic activities for DON-HRP/AuNPs/PAMAM were recorded compared with that of DON-HRP/AuNPs because DON-HRP/AuNPs/PAMAM presented higher relative catalytic activities. It was speculated that the outside numerous branches could provide the soft linker to combine with protein molecules and reduce spatial potential resistance, which contributed to the higher activities of the protein.

Through exploring the effect of physicochemical factors on nanoparticles in immunoassay, it was observed that AuNPs coating with PAMAM displayed higher tolerance towards high salt concentration and organic solution, compared with naked AuNPs, and further guaranteed the performance of the assays. Then we could confirm that dendrimers could assist nanoparticles to fight against the adverse factors in the environment, which would enhance the application of various nanoparticles in diagnosis, imaging, drug delivery and other fields.

After integrated with IMBs and DON-HRP/AuNPs/PAMAM, a novel competitive immunoassay was established, in which the LOD was improved 4-fold in comparison with that based on DON-HRP/AuNPs and 13-fold with that based on DON-HRP. It was suggested that multi-enzyme probes based on nanoparticles could promote the sensitivity of analytical methods with the aid of the higher specific surface area for the loading amount of HRP [18]. Next, the sensitivity of magnetic immunoassays based on DON-HRP/AuNPs/PAMAM was higher than that based on DON-HRP/AuNPs, which could be associated with the higher relative catalytic activity. In summary, dendrimer–inorganic NPs not only presented better stability but also could be undertaken as a friendly carrier to load bioactive macromolecules to play a major role in biomedicine, imaging and biosensors fields.

Various biosensors based on nanoparticles for the detection of DON have been reported in Table 3. Conventional immunosensors, such as lateral flow immunoassays or electrochemical immunoassays, were integrated with different nanostructures to establish the analysis methods. In electrochemical immunoassays, nanoparticles were adopted to modify the electrode, including hierarchical nanostructured CoCo Prussian blue analog entrapped by a Zr-based porphyrin MOF [39], multi-walled carbon nanotubes (MWCNTs) [40] and PtPd nanoparticles compound polyethyleneimine-functionalized reduced graphene oxide (PtPd NPs/PEIrGO) [41]. For lateral flow immunoassays, nanomaterials with different physicochemical properties served as signal tags, such as amorphous carbon nanoparticles (ACNPs) with strong dark color [42], fluorescent microsphere [43] and α-Fe_2O_3 nanocubes (FNCs) with orange color [44]. Along with the application of nanoparticles, these biosensors displayed higher sensitivity and the lowest detection limit was 0.14 fg/mL [39], implying the important role of nanomaterials in assays. Moreover, lateral flow immunoassays represented more advantages in simplicity, but the sensitivity needs to be improved. Electrochemical immunoassays displayed the highest sensitivity along with complex operation. In comparison, magnetic immunoassays based on DON-HRP/Au/PAMAM presented better performance than the majority of the listed assays [40–45].

Table 3. Nanoparticles-based biosensors for the detection of DON.

Nanoparticles	Biosensors	Recognition Molecule	LOD	Refs.
CoCoPBA@PCN-221 nanostructure	Electrochemical immunoassays	Antibody	0.14 fg/mL	[39]
MWCNTs	Electrochemical immunoassays	Molecularly imprinted poly(l-arginine)	0.07 μM	[40]
PtPd NPs/PEI-rGO	Electrochemical immunoassays	Aptamers	6.9 ng/mL	[41]
ACNPs	Lateral flow immunoassays	Antibody	20 ng/g	[42]
Fluorescent microsphere	Lateral flow immunoassays	Antibody	2.5 ng/mL	[43]
FNCs	Lateral flow immunoassays	Antibody	0.18 ng/mL	[44]
AuNPs	Lateral flow immunoassays	Antibody	12.5 ng/mL	[45]
Fe_3O_4@polydopamine	Microchannel resistance sensor	Antibody	20.7 pg/mL	[46]
AuNPs	Colorimetric biosensor	Antibody	0.127 ng/mL	This work
AuNPs/PAMAM	Colorimetric biosensor	Antibody	0.035 ng/mL	This work

It is important to evaluate the specificity of the immunosensors since various pollutants co-exist in real samples. The analogies displayed no obvious effects of interferences in the new method, including T-2, ZEN, OTA, AFB1 and AFM1. As biosynthetic precursors of DON, there was lower cross-reactivity with 3-Ac-DON and 15-Ac-DON because of the similar molecule structures. The results indicated that magnetic immunoassays based on DON-HRP/Au/PAMAM possessed good selectivity and specificity for DON detection.

5. Conclusions

In this study, AuNPs using PAMAM as a host were synthesized and characterized. Multiple enzymatic labels based on Au/PAMAM were developed and characterized. Antibodies were immobilized on IMBs through Protein G directionally. Magnetic immunoassays based on different enzyme complexes were established and compared. Magnetic immunoassays based on DON-HRP/Au/PAMAM presented the highest sensitivity with the LOD of 0.035 ng/mL, improving 13-fold compared with conventional magnetic immunoassays. The recovery of this new method for rice, corn and wheat was 90.8–116.2%. In real grain samples, the concentration of DON was ND-34.76 ng/g and the average detection rate was 43.33%. It was inspired that dendrimer–nanoparticles can yield novel strategies for universal amplification labels in biosensing and molecular diagnostics.

Supplementary Materials: The following supporting information can be downloaded at https://www.mdpi.com/article/10.3390/bios13050536/s1, Figure S1: TEM images of AuNPs; Figure S2: Correlation analysis between magnetic immunoassay based on DON-HRP/AuNPs/PAMAM and UPLC-MS/MS; Table S1: Optimization of parameters in magnetic immunoassays based on DON-HRP/AuNPs; Table S2: Optimization of parameters in magnetic immunoassays based on DON-HRP/AuNPs/PAMAM.

Author Contributions: Conceptualization, K.Z. and Z.Z.; methodology, J.Y. and H.S.; validation, formal analysis and data curation, S.Y. and X.G.; writing—original draft preparation, K.Z.; writing—review and editing, supervision and funding acquisition, Z.Z. and H.Z. All authors have read and agreed to the published version of the manuscript.

Funding: This work was supported financially by the Jiangsu Agricultural Science and Technology Independent Innovation Fund Project (CX (21)3173), Quzhou High-Level Medical and Health Talents Program (KYQD2022-26) and the Jiangsu Collaborative Innovation Center of Technology and Material of Water Treatment.

Institutional Review Board Statement: Not applicable.

Informed Consent Statement: Not applicable.

Data Availability Statement: The data that support the findings of this study are available upon reasonable request.

Conflicts of Interest: The authors declare no conflict of interest.

References

1. Brase, S.; Encinas, A.; Keck, J.; Nising, C.F. Chemistry and biology of mycotoxins and related fungal metabolites. *Chem. Rev.* **2009**, *109*, 3903–3990. [CrossRef] [PubMed]
2. Ran, R.; Wang, C.H.; Han, Z.; Wu, A.B.; Zhang, D.B.; Shi, J.X. Determination of deoxynivalenol (DON) and its derivatives: Current status of analytical methods. *Food Control* **2013**, *34*, 138–148. [CrossRef]
3. Collins, T.F.X.; Sprando, R.L.; Black, T.N.; Olejnik, N.; Eppley, R.M.; Hines, F.A.; Rorie, J.; Ruggles, D.I. Effects of deoxynivalenol (DON, vomitoxin) on in utero development in rats. *Food Chem. Toxicol.* **2006**, *44*, 747–757. [CrossRef] [PubMed]
4. Ren, Z.H.; Deng, Y.T.; Wang, Y.C.; Zhou, R.; Deng, H.D.; Deng, J.L.; Zuo, Z.C.; Peng, X.; Yu, S.M.; Shen, L.H. Effects of the Fusarium toxin zearalenone and/or deoxynivalenol on the serum IL-1, IL-4, and C3 levels in mice. *Food Agric. Immunol.* **2016**, *27*, 414–421. [CrossRef]
5. Pestka, J.J. Deoxynivalenol: Mechanisms of action, human exposure, and toxicological relevance. *Arch. Toxicol.* **2010**, *84*, 663–679. [CrossRef]
6. *Commission Regulation (EC) No 1881/2006 of 19 December 2006 Setting Maximum Levels for Certain Contaminants in Foodstuffs*; FAO: Rome, Italy, 2006; p. L364/5.
7. *GB 2761-2017*; National Food Safety Standard Maximum Levels of Mycotoxins in Foods . National Health and Family Planning Commission: Beijing, China; State Food and Drug Administration: Beijing, China, 2017.
8. Omurtag, G.Z.; Beyoglu, D. Occurrence of deoxynivalenol (vomitoxin) in beer in Turkey detected by HPLC. *Food Control* **2007**, *18*, 163–166. [CrossRef]

9. Ok, H.E.; Lee, S.Y.; Chun, H.S. Occurrence and simultaneous determination of nivalenol and deoxynivalenol in rice and bran by HPLC-UV detection and immunoaffinity cleanup. *Food Control* **2018**, *87*, 53–59. [CrossRef]
10. Vendl, O.; Berthiller, F.; Crews, C.; Krska, R. Simultaneous determination of deoxynivalenol, zearalenone, and their major masked metabolites in cereal-based food by LC-MS-MS. *Anal. Bioanal. Chem.* **2009**, *395*, 1347–1354. [CrossRef]
11. Deng, C.L.; Li, C.L.; Zhou, S.; Wang, X.D.; Xu, H.B.; Wang, D.; Gong, Y.Y.; Routledge, M.N.; Zhao, Y.F.; Wu, Y.N. Risk assessment of deoxynivalenol in high-risk area of China by human biomonitoring using an improved high throughput UPLC-MS/MS method. *Sci. Rep.* **2018**, *8*, 3901. [CrossRef]
12. Nan, M.N.; Xue, H.L.; Bi, Y. Contamination, detection and control of mycotoxins in fruits and vegetables. *Toxins* **2022**, *14*, 309. [CrossRef]
13. Hou, L.; Tang, Y.; Xu, M.D.; Gao, Z.Q.; Tang, D.P. Tyramine-based enzymatic conjugate repeats for ultrasensitive immunoassay accompanying tyramine signal amplification with enzymatic biocatalytic precipitation. *Anal. Chem.* **2014**, *86*, 8352–8358. [CrossRef] [PubMed]
14. Zhang, H.; Shi, Y.P.; Lan, F.; Pan, Y.; Lin, Y.K.; Lv, J.Z.; Zhu, Z.H.; Jiang, Q.; Yi, C. Detection of single-digit foodborne pathogens with the naked eye using carbon nanotube-based multiple cycle signal amplification. *Chem. Commun.* **2014**, *50*, 1848–1850. [CrossRef] [PubMed]
15. Li, D.Y.; Cui, Y.L.; Morisseau, C.; Gee, S.J.; Bever, C.S.; Liu, X.J.; Wu, J.; Hammock, B.D.; Ying, Y.B. Nanobody based immunoassay for human soluble epoxide hydrolase detection using polymeric horseradish peroxidase (PolyHRP) for signal enhancement: The rediscovery of PolyHRP? *Anal. Chem.* **2017**, *89*, 6249–6257. [CrossRef] [PubMed]
16. Xie, T.; Wang, A.M.; Huang, L.F.; Li, H.F.; Chen, Z.M.; Wang, Q.Y.; Yin, X.P. Recent advance in the support and technology used in enzyme immobilization. *Afr. J. Biotechnol.* **2009**, *8*, 4724–4733.
17. Jiang, L.; Wei, D.L.; Zeng, K.; Shao, J.; Zhu, F.; Du, D.L. An enhanced direct competitive immunoassay for the detection of kanamycin and tobramycin in milk using multienzyme-particle amplification. *Food Anal. Method.* **2018**, *11*, 2066–2075. [CrossRef]
18. Zeng, K.; Wei, D.L.; Zhang, Z.; Meng, H.; Huang, Z.; Zhang, X.Y. Enhanced competitive immunomagnetic beads assay with gold nanoparticles and carbon nanotube-assisted multiple enzyme probes. *Sensors Actuat. B Chem.* **2019**, *292*, 196–202. [CrossRef]
19. Zhang, Q.Z.; Zhao, B.; Yan, J.; Song, S.P.; Min, R.; Fan, C.H. Nanotube-based colorimetric probe for ultrasensitive detection of ataxia telangiectasia mutated protein. *Anal. Chem.* **2011**, *83*, 9191–9196. [CrossRef]
20. Bagwe, R.P.; Hilliard, L.R.; Tan, W. Surface modification of silica nanoparticles to reduce aggregation and nonspecific binding. *Langmuir* **2006**, *22*, 4357–4362. [CrossRef]
21. Frechet, J.M.J. Dendrimers and other dendritic macromolecules: From building blocks to functional assemblies in nanoscience and nanotechnology. *J. Polym. Sci. Pol. Chem.* **2003**, *41*, 3713–3725. [CrossRef]
22. Fernandes, T.; Daniel-Da-Silva, A.L.; Trindade, T. Metal-dendrimer hybrid nanomaterials for sensing applications. *Coordin. Chem. Rev.* **2022**, *460*, 214483. [CrossRef]
23. Bronstein, L.M.; Shifrina, Z.B. Dendrimers as encapsulating, stabilizing, or directing agents for inorganic nanoparticles. *Chem. Rev.* **2011**, *111*, 5301–5344. [CrossRef] [PubMed]
24. Yamamoto, K.; Imaoka, T.; Tanabe, M.; Kambe, T. New horizon of nanoparticle and cluster catalysis with dendrimers. *Chem. Rev.* **2020**, *120*, 1397–1437. [CrossRef] [PubMed]
25. Caminade, A.M.; Hameau, A.; Turrin, C.O.; Laurent, R.; Majoral, J.P. Dendritic metal complexes for bioimaging. Recent advances. *Coordin. Chem. Rev.* **2021**, *430*, 213739. [CrossRef]
26. Patle, R.Y.; Meshram, J.S. The advanced synthetic modifications and applications of multifunctional PAMAM dendritic composites. *React. Chem. Eng.* **2021**, *7*, 9–40. [CrossRef]
27. Viltres, H.; Lopez, Y.C.; Leyva, C.; Gupta, N.K.; Naranjo, A.G.; Acevedo-Pena, P.; Sanchez-Diaz, A.; Bae, J.; Kim, K.S. Polyamidoamine dendrimer-based materials for environmental applications: A review. *J. Mol. Liq.* **2021**, *334*, 116017. [CrossRef]
28. Xu, Y.; Huang, Z.B.; He, Q.H.; Deng, S.Z.; Li, L.S.; Li, Y.P. Development of an immunochromatographic strip test for the rapid detection of deoxynivalenol in wheat and maize. *Food Chem.* **2010**, *119*, 834–839. [CrossRef]
29. Wu, Q.Y.; Zhao, F.; Yang, H.F.; Chen, H.Z.; Zhuang, Y.Q. Simultaneous determination of 5 deoxynivalenols in wheat grain by ultra-high performance liquid chromatography-tandem mass spectrometry. *Chin. J. Anal. Lab.* **2020**, *39*, 715.
30. Welch, N.G.; Scoble, J.A.; Muir, B.W.; Pigram, P.J. Orientation and characterization of immobilized antibodies for improved immunoassays. *Biointerphases* **2017**, *12*, 02D301. [CrossRef]
31. Hao, X.; Yang, X.; Zou, S.; Cao, X. Surface modification of poly(styrene) 96-well plates using aptamers via a dendrimer-templated strategy to enhance ELISA performances. *Colloid Surf. B* **2023**, *221*, 113003. [CrossRef]
32. Liu, Y.; Yu, J. Oriented immobilization of proteins on solid supports for use in biosensors and biochips: A review. *Microchim. Acta* **2016**, *183*, 1–19. [CrossRef]
33. Glavan, A.C.; Niu, J.; Chen, Z.; Guder, F.; Cheng, C.M.; Liu, D.; Whitesides, G.M. Analytical Devices Based on Direct Synthesis of DNA on Paper. *Anal. Chem.* **2016**, *88*, 725–731. [CrossRef] [PubMed]
34. Guo, S.L.; Chen, P.C.; Chen, M.S.; Cheng, Y.C.; Lin, J.M.; Lee, H.C.; Chen, C.S. A fast universal immobilization of immunoglobulin G at 4 degrees C for the development of array-based immunoassays. *PLoS ONE* **2012**, *7*, e51370. [CrossRef] [PubMed]
35. Iijima, M.; Nakayama, T.; Kuroda, S. Two-dimensional membrane scaffold for the oriented immobilization of biosensing molecules. *Biosens. Bioelectron.* **2020**, *150*, 111860. [CrossRef] [PubMed]
36. Jansen, J.F.G.A.; Meijer, E.W.; de Brabander-van den Berg, E.M.M. The dendritic box: Shape-selective liberation of encapsulated guests. *J. Am. Chem. Soc.* **1995**, *117*, 4417–4418. [CrossRef]

37. Camarada, M.B. PAMAM Dendrimers as support for the synthesis of gold nanoparticles: Understanding the effect of the terminal groups. *J. Phys. Chem. A* **2017**, *121*, 8124–8135. [CrossRef]
38. Hu, L.; Dong, T.; Zhao, K.; Deng, A.; Li, J. Ultrasensitive electrochemiluminescent brombuterol immunoassay by applying a multiple signal amplification strategy based on a PAMAM-gold nanoparticle conjugate as the bioprobe and Ag@Au core shell nanoparticles as a substrate. *Microchim. Acta* **2017**, *184*, 3415–3423. [CrossRef]
39. Cui, J.; Wu, B.W.; Li, Z.Z.; Bai, Y.H.; Kan, L.; Wang, M.H.; He, L.H.; Du, M. Hierarchical CoCoPBA@PCN-221 nanostructure for the highly sensitive detection of deoxynivalenol in foodstuffs. *Food Chem.* **2023**, *403*, 134370. [CrossRef]
40. Li, W.; Diao, K.S.; Qiu, D.Y.; Zeng, Y.F.; Tang, K.J.; Zhu, Y.F.; Sheng, Y.Y.; Wen, Y.P.; Li, M.F. A highly-sensitive and selective antibody-like sensor based on molecularly imprinted poly(L-arginine) on COOH-MWCNTs for electrochemical recognition and detection of deoxynivalenol. *Food Chem.* **2021**, *350*, 129229. [CrossRef]
41. Wang, K.; He, B.S.; Xie, L.L.; Li, L.P.; Liu, R.L.; Wei, M.; Jin, H.L.; Ren, W.J. Exonuclease III-assisted triple-amplified electrochemical aptasensor based on PtPd NPs/PEI-rGO for deoxynivalenol detection. *Sens. Actuat. B Chem.* **2021**, *349*, 130767. [CrossRef]
42. Zhang, X.Y.; Yu, X.Z.; Wen, K.; Li, C.L.; Marti, G.M.; Jiang, H.Y.; Shi, W.M.; Shen, J.Z.; Wang, Z.H. Multiplex lateral flow immunoassays based on amorphous carbon nanoparticles for detecting three fusarium mycotoxins in maize. *J. Agric. Food Chem.* **2017**, *65*, 8063–8071. [CrossRef]
43. Zhou, S.; Xu, L.G.; Kuang, H.; Xiao, J.; Xu, C.L. Fluorescent microsphere immunochromatographic sensor for ultrasensitive monitoring deoxynivalenol in agricultural products. *Microchem. J.* **2021**, *164*, 106024. [CrossRef]
44. Zhao, S.; Bu, T.; He, K.Y.; Bai, F.; Zhang, M.; Tian, Y.M.; Sun, X.Y.; Wang, X.; Zhangsun, H.; Wang, L. A novel α-Fe2O3 nanocubes-based multiplex immunochromatographic assay for simultaneous detection of deoxynivalenol and aflatoxin B1 in food samples. *Food Control.* **2021**, *123*, 107811. [CrossRef]
45. Hou, S.; Ma, J.J.; Cheng, Y.Q.; Wang, H.G.; Sun, J.H.; Yan, Y.X. One-step rapid detection of fumonisin B1, dexyonivalenol and zearalenone in grains. *Food Control* **2020**, *117*, 107107. [CrossRef]
46. Peng, X.; Dong, Y.Z.; Feng, N.; Wei, Q.L.; Lu, P.; Chen, Y.P. Fe_3O_4 @polydopamine-based microchannel resistance immunosensor for detecting deoxynivalenol in wheat samples. *Sensos Actuat. B Chem.* **2023**, *378*, 133151. [CrossRef]

 *biosensors* 

Communication

A Comparison Study of the Detection Limit of Omicron SARS-CoV-2 Nucleocapsid by Various Rapid Antigen Tests

Daniela Dobrynin [1], Iryna Polischuk [1] and Boaz Pokroy [1,2,*]

[1] Department of Materials Science and Engineering, Technion—Israel Institute of Technology, Haifa 32000, Israel
[2] The Russel Berrie Nanotechnology Institute, Technion—Israel Institute of Technology, Haifa 32000, Israel
* Correspondence: bpokroy@technion.ac.il

Abstract: Rapid antigen tests (RATs) are widely used worldwide to detect SARS-CoV-2 since they are an easy-to-use kit and offer rapid results. The RAT detects the presence of the nucleocapsid protein, which is located inside the virus. However, the sensitivity of the different RATs varies between commercially available kits. The test result might change due to various factors, such as the variant type, infection date, swab's surface, the manner in which one performs the testing and the mucus components. Here, we compare the detection limit of seven commercially available RATs by introducing them to known SARS-CoV-2 nucleocapsid protein amounts from the Omicron variant. It allows us to determine the detection limit, disregarding the influences of other factors. A lower detection limit of the RAT is necessary since earlier detection will help reduce the spread of the virus and allow faster treatment, which might be crucial for the population at risk.

Keywords: rapid antigen test; SARS-CoV-2; COVID-19; omicron variant

Citation: Dobrynin, D.; Polischuk, I.; Pokroy, B. A Comparison Study of the Detection Limit of Omicron SARS-CoV-2 Nucleocapsid by Various Rapid Antigen Tests. *Biosensors* **2022**, *12*, 1083. https://doi.org/10.3390/bios12121083

Received: 25 October 2022
Accepted: 24 November 2022
Published: 27 November 2022

1. Introduction

Since the first case of COVID-19 disease in Wuhan in December 2019, there has been a worldwide struggle to reduce the transmission of acute respiratory syndrome coronavirus 2 (SARS-CoV-2). Many countries worldwide imposed temporary local lockdowns in order to reduce person-to-person interactions [1,2], masks became obligatory, especially in closed spaces [3,4], and there was a general requirement for social distancing. However, the most efficient method to reduce the continuing spread of infection among the population, and in the meantime maintain a regular daily life, is the early detection of contagious infected people.

An extensive body of research aims to develop new technologies for SARS-CoV-2 detection [5–8]. One known and reliable method is the reverse-transcriptase PCR test (RT-PCR), which is constantly improving [9]. It is possible to detect the virus even if there is only one RNA strand in the sample and run hundreds of samples simultaneously [10]. This method has a few disadvantages, such as a high cost, is time consuming, the need for medical laboratories and skilled staff to perform the test, and the major flaw: the lack of an appropriate number of available tests. The latterly prominent Omicron variant (B.1.1.529) and its derivatives have caused a tremendous increase in the number of infected people due to its enhanced transmissibility [11,12]. All the above points emphasize the high demand for easy-to-use, cheap, and available detection tests for SARS-CoV-2 and other pathogens [13].

The solution was found in developing rapid antigen tests (RATs), which are based on lateral flow immunoassay technology (LFIA) and provide a result within 15–30 min. For a more detailed description regarding the principles of LFIA, readers are referred to Koczula et al. [14]. The majority of industrially manufactured RATs are designed based on the detection of the nucleocapsid protein, one of four main structural proteins in the SARS-CoV-2 [15]. The nucleocapsid protein has proved to be a good diagnostic marker since it is detectable even after one day from the disease onset [16]. The key role of the nucleocapsid protein is to pack the viral genome into helical complexes named ribonucleocapsid (RNP), which interacts with the membrane protein (M) [17].

Former studies tested the RATs' validity by examining PCR-positive COVID-19 patients samples or the isolated virus with various RATs [18–22]. These studies are very informative; however, many factors might affect the RAT result, such as the diversity in the viral load (VL, the virus amount in the sample) or the nucleocapsid concentration in each sample, the swab's surface, exposure to different variants, date of infection, immune system reaction to the virus, vaccination status, etc. For instance, it was shown that various RATs tend to detect the Delta or Omicron variants at different sensitivities [21,23]. The latter implies that infected patients who show PCR-positive results might have a RAT-negative result simply due to each person's uniqueness and situation. Moreover, the nasopharyngeal samples contain many other proteins which might affect the detection of the nucleocapsid protein. In previous studies, it was shown that there is a significant variation in sensitivity between different RATs when applied to both symptomatic participants (34.3–91.3%) and asymptomatic participants (28.6–77.8%) [24]. The sensitivity of RT-PCR tests is higher than that of the RAT tests [24,25]. Nevertheless, the lower the RAT's detection limit, the earlier one can detect COVID-19-infected patients. This is important in cases when urgent treatment is needed, as well as to facilitate early isolation in order not to infect others.

This study compares seven commercially available RATs for detecting SARS-CoV-2 by a simple direct experiment. In contrast to previous studies, here we examined the various RATs by testing with various known, pure, and pre-defined concentrations of Omicron SARS-CoV-2 nucleocapsid protein solutions. The tests were performed according to each manufacturer's instructions utilizing the RAT solution containing known volumes and concentrations of the nucleocapsid protein of the Omicron variant. This method allowed us to determine the detection limit of each RAT while controlling the amount of the added nucleocapsid protein and disregarding all other influencing factors.

2. Materials and Methods

The nucleocapsid protein of the Omicron variant was purchased from BPS Bioscience (San Diego, CA, USA) and contained the mutations: P13L, R203K, G204R, and Δ31-33ERSdel. We used seven commercially available antigen tests: (a) Deepblue COVID-19 (SARS-CoV-2) Antigen Test Kit (colloidal gold) (Anhui Deepblue Medical Technology Co., Ltd., Hefei, Anhui, China); (b) Easy Diagnosis COVID-19 (SARS-CoV-2) Antigen Test Kit (Wuhan EasyDiagnosis Biomedicine Co., Ltd., Wuhan, Hubei, China); (c) EcoTest COVID-19 TO-GO (Assure Tech. (Hangzhou) Co., Ltd., Hangzhou, China); (d) GenSure COVID-19 Antigen Rapid Test Kit (GenSure Biotech Inc., Shijiazhuang, Hebei, China); (e) Orient Gene Rapid COVID-19 (Antigen) Self-Test (Zhejiang Orient Gene Biotech Co., Ltd., Huzhou, Zhejiang, China); (f) Panbio COVID-19 Antigen Self-Test (Abbott Laboratories, Chicago, IL, USA); (g) YHLO GLINE-2019-nCoV Ag for self-testing (Shenzhen YHLO Biotech Co., Ltd., Shenzhen, China).

The protein was diluted in Phosphate buffer saline (PBS), pH 7.4 (Sigma-Aldrich, St. Louis, MO, USA) to various concentrations: 550; 55; 5.5; 2.75; 1; 0.55; and 0.275 μg mL^{-1}. After use, the protein solutions were stored at −20 °C (for the experiments which were conducted in March and April 2022). For the experiments held in October 2022, new protein solutions were prepared in the same way as before (the same pipettes and volumes were used). The (f) RAT is not included in these experiments (RH 60%) since, by this time, the RAT was not marketing anymore. The highest concentration is 550 μg mL^{-1}, similar to the concentration delivered by BPS Bioscience (580 μg mL^{-1}). A total of 5 μL from the relevant nucleocapsid concentration solution was added to each RAT test solution with a pipette. There was no use of a swab. The volume of the various test solutions varies between the different companies (250–400 μL, Table S1), but the total protein amount present in each test solution is identical. The protein in the test solution was mixed by pipetting until the protein was homogeneously dissolved. At the next step, as described in each manufacturer's instructions, a specific number of drops were dripped from the test solution to the designated sample loading location at the RAT. After 15 min, the tested RATs were examined by the naked eye, and images were acquired (Figures S1–S7 in Supplementary Materials).

Performance of all the RATs is based on the same concept; the C-line is the control line and should appear if the RAT was conducted correctly, and the T-line appears if the nucleocapsid protein was detected.

Each type of RAT was firstly checked with the highest nucleocapsid concentration, and if the result was clearly positive (the T-line was bold and clear), we proceeded to test lower concentrations. If the T-line was barely visible, we performed three RAT replicates and continued to lower the protein concentration. Once the T-line was completely not visible, we again tested three replicates to ensure a negative result. The detection limit was determined as the lowest protein concentration, where all three replicates show a positive result.

All experiments were conducted in the fume hood at 21 °C. The relative humidity (RH) was stabilized by dehumidifier SAD50 (SmartAir, Kfar Kasem, Israel) to 30% (March and April months) and 60% (October).

3. Results and Discussion

This work provides an evaluation of seven commercially available RATs widely used internationally. In order to determine and compare the detection limit of different RATs, we directly used Omicron nucleocapsid protein solutions in decreasing concentrations. The protein concentration was lowered continuously until only the control line (C-line) was visible, rather than the two lines (C- and T-lines). The same volume of the protein solution (5 μL) was added to the test solutions each time, implying that the total amount of the nucleocapsid was identical. The performance of the different RATs was further compared, eliminating any human factor, such as the manner in which one may conduct the RAT and the actual viral load of the tested patient. The experiments were conducted under different relative humidity (RH) conditions of 30% and 60%.

Our results suggest that all RATs were able to detect the nucleocapsid protein using the 5.5 μg mL^{-1} solution, while the lowest detected concentration was 0.55 μg mL^{-1} by (b) RAT (Figure 1). In general, the reduction in the concentration of nucleocapsid protein caused the T-line to be weaker and thinner, as expected. At RH 60%, (c), (d), and (e) RATs were able to detect the nucleocapsid until 1 μg mL^{-1} solution. (g) RAT showed positive results until 2.75 μg mL^{-1} nucleocapsid concentration. These results provide additional evidence to previous studies suggesting that there might be significant variation between different RATs in detecting SARS-CoV-2, especially the Omicron variant. In this study, we observe up to one order of magnitude of nucleocapsid concentration difference between the several RATs. A reduced sensitivity was observed at RH 30% compared to RH 60% for the same protein concentration. For instance, for (b) RAT at RH 30%, only one out of three replicates showed a positive result at a concentration of 0.55 μg mL^{-1}, while at RH 60% all replicates showed a clear T-line. In the case of the (c) and (d) RATs, the results are improved as well; for (c) RAT, all replicates showed positive results in RH 60% (1 μg mL^{-1}), while in RH 30%, only two out of three. The (d) RAT could not detect the protein in 0.55 μg mL^{-1} concentration in RH 30%, in contrast to RH 60%, where two out of three replicates showed a T-line. A previous study showed that at an RH higher than 60% the water evaporation is reduced, and the diffusion across the nitrocellulose strip is improved, which obtains the desired signal [26]. The different results in the same RATs might be due to inaccuracies in the fabrication process or storage and shipment conditions.

There might be several explanations for the change in the detection limit between the different RATs, which are mainly attributed to variations in the manufacturing process. For instance, the T-line color stems from various sources, in particular either colloidal gold (RATs (a), (b), (g)), color conjugation (RAT(c)), polymer label (RAT (d)), etc. Additionally, the development process of the anti-nucleocapsid antibodies might vary between the companies, which would cause the recognition of different epitopes with different affinity and specificity. Moreover, the number of conjugated antibodies at the T-line location might also affect the detection limit. The latter might be the cause for achieving different results under the same conditions (same RAT, concentration, and RH).

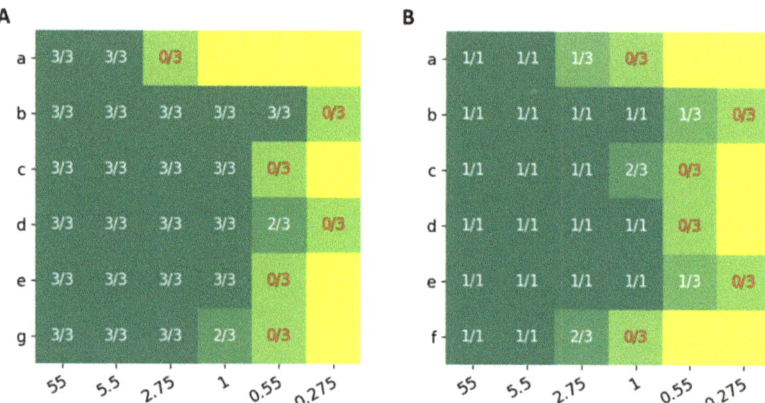

Figure 1. Various RATs were tested to detect a pure nucleocapsid protein in different concentrations, at RH conditions of (**A**) 60% and (**B**) 30%. The RATs are: (a) Deepblue COVID-19 (SARS-CoV-2) Antigen Test Kit (colloidal gold); (b) Easy Diagnosis COVID-19 (SARS-CoV-2) Antigen Test Kit; (c) EcoTest COVID-19 TO-GO; (d) GenSure COVID-19 Antigen Rapid Test Kit; (e) Orient Gene Rapid COVID-19 (Antigen) Self-Test; (f) Panbio COVID-19 Antigen Self-Test; (g) YHLO GLINE-2019-nCoV Ag for self-testing. Within each square are presented a number of positive results out of the replicates' number.

As seen in Figure S8, we compared three different RATs using the same nucleocapsid concentration of the nucleocapsid protein. The emerged T-lines are not similar; in the case of RAT (f), the T-line is very thick and clear, the T-line observed on RAT (d) is weaker but detectable by the human eye, and in the case of RAT (a), the T-line is the least prominent. On the other hand, RAT (d) shows a better detection capacity than these two RATs.

Previous studies have shown that all twelve nucleocapsid proteins interact with ~800 nucleotides of genomic RNA to produce an RNP complex [27]. Within each virus, there are 35–40 RNPs, meaning that the total amount of nucleocapsid protein varies between 420 and 480 entities according to this model. Assuming that the weight of the protein is 48 kDa, we can translate the amount of protein to the corresponding number of viruses. For instance, 5 μL of 55 μg mL^{-1} nucleocapsid solution is equivalent to a protein mass of 275 ng, which translates to 7.19–8.21 $\times$ 10^9 virions. The lowest detectable protein concentration of 0.55 μg mL^{-1} corresponds to 7.19–8.21 $\times$ 10^7 virions. This analysis emphasizes the high amount of virus copies needed to receive a positive result, in contrast to nucleic acid amplification techniques, which can detect an infected person in the presence of only several virions.

The average VL of nasopharyngeal swabs (NS) in infected patients is about ~10^8 genome copies per mL, and it varies between 10^5–10^{12} genome copies per mL [28,29]. Such high variation is possible due to various factors, such as vaccination, variant type, number of days since infection, etc. For instance, the VL of vaccinated patients is lower than that of non-vaccinated patients, and the Delta variant causes a higher VL than that of the Omicron variant [29]. Thus, it is reasonable that a higher VL will lead to a higher amount of the nucleocapsid protein in the test solution and a lower probability of having a false-negative result.

Our study has several limitations. We have consistently added a specific volume of the protein solution to the test solution. In contrast, during real-life testing, the nasal mucus that discharges to the test solution is not identical and might be higher or lower. Moreover, here we use pure nucleocapsid protein dissolved in PBS, while the nasal samples contain the whole virus, and the nucleocapsid interacts with the RNA and M protein of the virus. This might cause the nucleocapsid epitopes to be unavailable for interaction with the RAT's antibodies and reduce the detection efficiency. Furthermore, the nasal samples contain various additional proteins, such as mucin, and even other microorganisms. The factors mentioned above might cause changes in the interaction of the nucleocapsid protein with the antibodies present in the RAT, in contrast to the pure nucleocapsid we used. However, the experimental conditions are identical for all the tested RATs, which allows a reliable comparison. In addition, as we can see in Figures S1–S7, for the lowest detectable protein concentration, in some cases only, one out of three replicates showed a positive result. This means there are also variations in performance between the RATs from the same company. The increased number of tested RATs of each type may provide an even more reliable determination of the detection limits. However, the question arises whether the patient will persist and perform the test in a real-life situation more than a few times.

4. Conclusions

Our study demonstrates a simple method to evaluate the detection limit of various RATs using known pre-defined amounts of nucleocapsid protein. Here, as previously shown in other studies, yet in a new manner, there is a noticeable difference in the detection limits of various commercially available RATs, which customers and medical workers need to take into account when performing the RAT. We found the nucleocapsid protein mass of the various commercial RATs in the range of 2.75–27.5 ng in a test solution to be the lower detection limit for the RATs. This protein mass correlates to protein concentrations in the test solutions ranging between 9.8–78.6 ng mL^{-1}. Obviously, this demonstrates not only the detection limit of RATs in general but also the variation in detection limits between various commercial RATs (eight-fold). Using a pure recombinant protein for RATs' assessment is a simple and cheap method and does not require performing tests on patients (i.e., avoiding the need for a Helsinki declaration) or preserving their samples. For future use, if a new variant of concern emerges, for a quick assessment, it would not be necessary to wait for infected patients in order to test the RATs on human samples, but rather study the updated recombinant protein and compare it to previous ones. Additionally, it will allow the RATs producers to evaluate whether any change is needed in previously marketed RATs, to obtain more reliable tests.

Supplementary Materials: The following supporting information can be downloaded at: https://www.mdpi.com/article/10.3390/bios12121083/s1, Table S1. Volumes of test solutions of various rapid antigen tests; Figure S1. "Deepblue COVID-19 (SARS-CoV-2) Antigen Test Kit (colloidal gold)" RATs after reaction with different concentrations of nucleocapsid protein at RH conditions of (A) 30% and (B) 60%. The lowest concentration that showed a positive result is 2.75 μg mL^{-1}; only one out of three RATs showed a weak T-line (red arrow) at RH 30%; Figure S2. "Easy Diagnosis COVID-19 (SARS-CoV-2) Antigen Test Kit" RATs after reaction with different concentrations of nucleocapsid protein at RH conditions of (A) 30% and (B) 60%. The lowest concentration that showed positive result, in all replicates at RH 60%, is 0.55 μg mL^{-1}; only one out of three RATs showed a weak T-line (red arrow) at RH 30%; Figure S3. "EcoTest COVID-19 TO-GO" RATs after reaction with different concentrations of nucleocapsid protein. The lowest concentration that showed positive result is 1 μg mL^{-1} at RH conditions of (A) 30% and (B) 60%; Two out of three RATs showed a weak T-line (red arrows) at RH 30% and all replicates at 60%; Figure S4. "GenSure COVID-19 Antigen Rapid Test Kit" RATs after reaction with different concentrations of nucleocapsid protein at RH conditions of (A) 30% and (B) 60%. The lowest concentration that showed positive result is 0.55 μg mL^{-1} at RH 60%; two out of three tests showed a positive result (red arrows); Figure S5. "Orient Gene Rapid COVID-19 (Antigen) Self-Test" RATs after reaction with different concentrations of nucleocapsid protein at RH conditions of (A) 30% and (B) 60%. The lowest concentration that showed a positive result is 1 μg mL^{-1} at RH 60%. Only one out of three RATs showed a weak T-line (red arrow) using a concentration of 0.55 μg mL^{-1}, RH 30%; Figure S6. "Panbio COVID-19 Antigen Self-Test" RATs after reaction with different concentrations of nucleocapsid protein at RH 30%. The lowest concentration that showed positive result is 2.75 μg mL^{-1}; Two out of three RATs showed a T-line (red arrows); Figure S7. "YHLO GLINE-2019-nCoV Ag for self-testing" RATs after reaction with different concentrations of nucleocapsid protein at RH 60%. The lowest concentration that showed positive result is 1 μg mL^{-1}; Two out of three RATs showed a weak T-line (red arrow); Figure S8. Three different RATs after reaction with 5.5 μg mL^{-1} nucleocapsid concentration on the same day. The T-line is the least prominent in the case of "Deepblue COVID-19 (SARS-CoV-2) Antigen Test Kit (colloidal gold)" RAT, intermediate intensity of the T-line observed in the case of "GenSure COVID-19 Antigen Rapid Test Kit" RAT, and the "Panbio COVID-19 Antigen Self-Test" RAT showed the most prominent T-line.

Author Contributions: Conceptualization, B.P.; methodology, D.D. and I.P.; writing—original draft preparation, D.D.; writing—review and editing, I.P. and B.P.; supervision, B.P.; funding acquisition, B.P. All authors have read and agreed to the published version of the manuscript.

Funding: This research was funded by the Israel Science Foundation (ISF Grant 3970/19).

Institutional Review Board Statement: Not applicable.

Informed Consent Statement: Not applicable.

Data Availability Statement: Not applicable.

Acknowledgments: B.P. thanks the Bank Discount Academic Chair for financial support. B.P. thanks A. Futerman from The Weizmann Institute of Science.

Conflicts of Interest: The authors declare no conflict of interest.

References

1. The Lancet. India under COVID-19 Lockdown. *Lancet* **2020**, *395*, 1315. [CrossRef]
2. Di Domenico, L.; Pullano, G.; Sabbatini, C.E.; Boëlle, P.-Y.; Colizza, V. Impact of Lockdown on COVID-19 Epidemic in Île-de-France and Possible Exit Strategies. *BMC Med.* **2020**, *18*, 240. [CrossRef] [PubMed]
3. Feng, S.; Shen, C.; Xia, N.; Song, W.; Fan, M.; Cowling, B.J. Rational Use of Face Masks in the COVID-19 Pandemic. *Lancet Respir. Med.* **2020**, *8*, 434–436. [CrossRef] [PubMed]
4. Chua, M.H.; Cheng, W.; Goh, S.S.; Kong, J.; Li, B.; Lim, J.Y.C.; Mao, L.; Wang, S.; Xue, K.; Yang, L.; et al. Face Masks in the New COVID-19 Normal: Materials, Testing, and Perspectives. *Research* **2020**, *2020*, 7286735. [CrossRef] [PubMed]
5. Erdem, Ö.; Eş, I.; Saylan, Y.; Inci, F. Unifying the Efforts of Medicine, Chemistry, and Engineering in Biosensing Technologies to Tackle the Challenges of the COVID-19 Pandemic. *Anal. Chem.* **2022**, *94*, 3–25. [CrossRef]
6. Huergo, M.A.C.; Thanh, N.T.K. Current Advances in the Detection of COVID-19 and Evaluation of the Humoral Response. *Analyst* **2021**, *146*, 382–402. [CrossRef]
7. Thapa, S.; Singh, K.R.; Verma, R.; Singh, J.; Singh, R.P. State-of-the-Art Smart and Intelligent Nanobiosensors for SARS-CoV-2 Diagnosis. *Biosensors* **2022**, *12*, 637. [CrossRef]
8. Asghar, R.; Rasheed, M.; ul Hassan, J.; Rafique, M.; Khan, M.; Deng, Y. Advancements in Testing Strategies for COVID-19. *Biosensors* **2022**, *12*, 410. [CrossRef]
9. Wang, J.; Cai, K.; Zhang, R.; He, X.; Shen, X.; Liu, J.; Xu, J.; Qiu, F.; Lei, W.; Wang, J.; et al. Novel One-Step Single-Tube Nested Quantitative Real-Time PCR Assay for Highly Sensitive Detection of SARS-CoV-2. *Anal. Chem.* **2020**, *92*, 9399–9404. [CrossRef]
10. Guglielmi, G. The Explosion of New Coronavirus Tests That Could Help to End the Pandemic. *Nature* **2020**, *583*, 506–509. [CrossRef]
11. Chen, J.; Wang, R.; Gilby, N.B.; Wei, G.-W. Omicron Variant (B.1.1.529): Infectivity, Vaccine Breakthrough, and Antibody Resistance. *J. Chem. Inf. Model.* **2022**, *62*, 412–422. [CrossRef] [PubMed]
12. Poudel, S.; Ishak, A.; Perez-Fernandez, J.; Garcia, E.; León-Figueroa, D.A.; Romaní, L.; Bonilla-Aldana, D.K.; Rodriguez-Morales, A.J. Highly Mutated SARS-CoV-2 Omicron Variant Sparks Significant Concern among Global Experts—What Is Known so Far? *Travel Med. Infect. Dis.* **2022**, *45*, 102234. [CrossRef] [PubMed]
13. Castillo-León, J.; Trebbien, R.; Castillo, J.J.; Svendsen, W.E. Commercially Available Rapid Diagnostic Tests for the Detection of High Priority Pathogens: Status and Challenges. *Analyst* **2021**, *146*, 3750–3776. [CrossRef]
14. Koczula, K.M.; Gallotta, A. Lateral Flow Assays. *Essays Biochem.* **2016**, *60*, 111–120. [CrossRef] [PubMed]
15. Bai, Z.; Cao, Y.; Liu, W.; Li, J. The SARS-CoV-2 Nucleocapsid Protein and Its Role in Viral Structure, Biological Functions, and a Potential Target for Drug or Vaccine Mitigation. *Viruses* **2021**, *13*, 1115. [CrossRef]
16. Che, X.-Y.; Hao, W.; Wang, Y.; Di, B.; Yin, K.; Xu, Y.-C.; Feng, C.-S.; Wan, Z.-Y.; Cheng, V.C.C.; Yuen, K.-Y. Nucleocapsid Protein as Early Diagnostic Marker for SARS. *Emerg. Infect. Dis.* **2004**, *10*, 1947–1949. [CrossRef] [PubMed]

17. Lu, S.; Ye, Q.; Singh, D.; Cao, Y.; Diedrich, J.K.; Yates, J.R.; Villa, E.; Cleveland, D.W.; Corbett, K.D. The SARS-CoV-2 Nucleocapsid Phosphoprotein Forms Mutually Exclusive Condensates with RNA and the Membrane-Associated M Protein. *Nat. Commun.* **2021**, *12*, 502. [CrossRef] [PubMed]
18. Diao, B.; Wen, K.; Zhang, J.; Chen, J.; Han, C.; Chen, Y.; Wang, S.; Deng, G.; Zhou, H.; Wu, Y. Accuracy of a Nucleocapsid Protein Antigen Rapid Test in the Diagnosis of SARS-CoV-2 Infection. *Clin. Microbiol. Infect.* **2021**, *27*, 289.e1–289.e4. [CrossRef] [PubMed]
19. Mak, G.C.; Cheng, P.K.; Lau, S.S.; Wong, K.K.; Lau, C.; Lam, E.T.; Chan, R.C.; Tsang, D.N. Evaluation of Rapid Antigen Test for Detection of SARS-CoV-2 Virus. *J. Clin. Virol.* **2020**, *129*, 104500. [CrossRef]
20. Porte, L.; Legarraga, P.; Vollrath, V.; Aguilera, X.; Munita, J.M.; Araos, R.; Pizarro, G.; Vial, P.; Iruretagoyena, M.; Dittrich, S.; et al. Evaluation of a Novel Antigen-Based Rapid Detection Test for the Diagnosis of SARS-CoV-2 in Respiratory Samples. *Int. J. Infect. Dis.* **2020**, *99*, 328–333. [CrossRef]
21. Osterman, A.; Badell, I.; Basara, E.; Stern, M.; Kriesel, F.; Eletreby, M.; Öztan, G.N.; Huber, M.; Autenrieth, H.; Knabe, R.; et al. Impaired Detection of Omicron by SARS-CoV-2 Rapid Antigen Tests. *Med. Microbiol. Immunol.* **2022**, *2022*, 730. [CrossRef] [PubMed]
22. Cubas-Atienzar, A.I.; Kontogianni, K.; Edwards, T.; Wooding, D.; Buist, K.; Thompson, C.R.; Williams, C.T.; Patterson, E.I.; Hughes, G.L.; Baldwin, L.; et al. Limit of Detection in Different Matrices of 19 Commercially Available Rapid Antigen Tests for the Detection of SARS-CoV-2. *Sci. Rep.* **2021**, *11*, 18313. [CrossRef] [PubMed]
23. Stanley, S.; Hamel, D.J.; Wolf, I.D.; Riedel, S.; Dutta, S.; Contreras, E.; Callahan, C.J.; Cheng, A.; Arnaout, R.; Kirby, J.E.; et al. Limit of Detection for Rapid Antigen Testing of the SARS-CoV-2 Omicron and Delta Variants of Concern Using Live-Virus Culture. *J. Clin. Microbiol.* **2022**, *60*, 22. [CrossRef]
24. Dinnes, J.; Sharma, P.; Berhane, S.; van Wyk, S.S.; Nyaaba, N.; Domen, J.; Taylor, M.; Cunningham, J.; Davenport, C.; Dittrich, S.; et al. Rapid, Point-of-Care Antigen Tests for Diagnosis of SARS-CoV-2 Infection. *Cochrane Database Syst. Rev.* **2022**, *2022*, CD013705. [CrossRef]
25. Chu, V.T.; Schwartz, N.G.; Donnelly, M.A.P.; Chuey, M.R.; Soto, R.; Yousaf, A.R.; Schmitt-Matzen, E.N.; Sleweon, S.; Ruffin, J.; Thornburg, N.; et al. Comparison of Home Antigen Testing with RT-PCR and Viral Culture during the Course of SARS-CoV-2 Infection. *JAMA Intern. Med.* **2022**, *182*, 701–709. [CrossRef]
26. Choi, J.R.; Hu, J.; Feng, S.; Wan Abas, W.A.B.; Pingguan-Murphy, B.; Xu, F. Sensitive Biomolecule Detection in Lateral Flow Assay with a Portable Temperature–Humidity Control Device. *Biosens. Bioelectron.* **2016**, *79*, 98–107. [CrossRef]
27. Klein, S.; Cortese, M.; Winter, S.L.; Wachsmuth-Melm, M.; Neufeldt, C.J.; Cerikan, B.; Stanifer, M.L.; Boulant, S.; Bartenschlager, R.; Chlanda, P. SARS-CoV-2 Structure and Replication Characterized by in Situ Cryo-Electron Tomography. *Nat. Commun.* **2020**, *11*, 5885. [CrossRef] [PubMed]
28. Singanayagam, A.; Hakki, S.; Dunning, J.; Madon, K.J.; Crone, M.A.; Koycheva, A.; Derqui-Fernandez, N.; Barnett, J.L.; Whitfield, M.G.; Varro, R.; et al. Community Transmission and Viral Load Kinetics of the SARS-CoV-2 Delta (B.1.617.2) Variant in Vaccinated and Unvaccinated Individuals in the UK: A Prospective, Longitudinal, Cohort Study. *Lancet Infect. Dis.* **2022**, *22*, 183–195. [CrossRef]
29. Puhach, O.; Adea, K.; Hulo, N.; Sattonnet, P.; Genecand, C.; Iten, A.; Bausch, F.J.; Kaiser, L.; Vetter, P.; Eckerle, I.; et al. Infectious Viral Load in Unvaccinated and Vaccinated Individuals Infected with Ancestral, Delta or Omicron SARS-CoV-2. *Nat. Med.* **2022**, *2022*, 816. [CrossRef]

MDPI

Article

Two-Way Detection of COVID-19 Spike Protein and Antibody Using All-Dielectric Metasurface Fluorescence Sensors

Masanobu Iwanaga * and Wanida Tangkawsakul

Research Center of Functional Materials, National Institute for Materials Science (NIMS), 1-1 Namiki, Tsukuba 305-0044, Japan

* Correspondence: iwanaga.masanobu@nims.go.jp

Abstract: COVID-19 (or SARS-CoV-2) has deeply affected human beings worldwide for over two years, and its flexible mutations indicate the unlikeliness of its termination in a short time. Therefore, it is important to develop a quantitative platform for direct COVID-19 detection and human status monitoring. Such a platform should be simpler than nucleic acid amplification techniques such as polymerase chain reaction, and more reliable than the disposable test kits that are based on immunochromatography. To fulfill these requirements, we conducted proof-of-concept experiments for the quantitative detection of spike glycoprotein peptides and antibodies in one platform, i.e., all-dielectric metasurface fluorescence (FL) sensors. The high capability to enhance FL intensity enabled us to quantitatively measure the glycoproteins and antibodies more efficiently compared with the previous methods reported to date. Furthermore, the intrinsic limit of detection in the metasurface FL sensors was examined via confocal microscopy and found to be less than 0.64 pg/mL for glycoprotein peptides. Moreover, the sensors had a dynamic range more than five orders that of the target concentrations, indicating extremely high sensitivity. These two-way functions of the metasurface FL sensors can be helpful in reducing daily loads in clinics and in providing quantitative test values for proper diagnosis and cures.

Keywords: COVID-19; SARS-CoV-2; spike glycoprotein; peptide; antibody; all-dielectric metasurface; fluorescence sensor; sandwich assay

Citation: Iwanaga, M.; Tangkawsakul, W. Two-Way Detection of COVID-19 Spike Protein and Antibody Using All-Dielectric Metasurface Fluorescence Sensors. *Biosensors* **2022**, *12*, 981. https://doi.org/10.3390/bios12110981

Received: 28 September 2022
Accepted: 2 November 2022
Published: 7 November 2022

1. Introduction

For more than two years, the COVID-19 (SARS-CoV-2) pandemic has substantially reduced social and economic activities. To overcome such damages, efficient and quantitative techniques for evaluating the status of human bodies are required. Particularly, quantification of the concentrations of COVID-19 and neutralizing antibodies (Abs) are highly preferred for proper therapeutic approaches. Currently, polymerase chain reaction (PCR) [1] serves as a high sensitivity method for COVID-19 detection; however, it is widely known that PCR is costly as a daily practice and has high demand with regard to human resources. As handy chips, antigen test kits relying on paper-based immunochromatography are used; they are known to be far less sensitive than PCR and qualitative platforms for supplementary tests. Numerous proposals for detecting COVID-19 have been reported in the last two years, and are summarized in several review papers [2–8]. However, to the best of our knowledge, there is no newly established technique that meets the aforementioned criterion, suggesting practical difficulties that have not been noted in previous reports. Thus, there is room to explore a practical and quantitative technique to detect the antigens and Abs.

Here, we address proof-of-concept experiments for the quantitative detection of COVID-19 spike proteins and Abs. When implementing sandwich assays to detect Ab–protein–Ab complexes, it becomes possible to set the proteins or Abs to be a target in the assay, meaning that two-way detection can be realized in one platform through the employment of all-dielectric metasurface fluorescence (FL) sensors [9–11]. The configurations of the two-way assay are schematically illustrated in Figure 1a,b. The optical features of the metasurfaces are described in Section 3.1. Section 3.2 shows that efficient detection of the target glycoprotein can be realized in a practical setup which is superior to the conventional enzyme-linked immunosorbent assays (ELISAs), and that the intrinsic detection capability of metasurface FL sensors reaches sub-pg/mL range in low-background confocal measurements. Additionally, we obtained linear quantitative detection down to 10 picomolars (pM) when setting the Abs as a target; this is elaborated in Section 3.3. Importantly, the all-dielectric metasurface FL sensors were applied to the detection of immunoglobulin G (IgG) Abs in human serums, and it was substantiated that they can retain their high sensitivity even in the presence of abundant other biomolecules, such as albumin, lipids, and other Abs [9]. One of the key factors is binding molecules that are immobilized on the surface of Si nanorods, which specifically capture biomolecules. In this study, we used the same binding molecule used in previous reports [9–11]. The metasurfaces were arrayed on a substrate in an assembled manner, as shown in Figure 1c. By combining the metasurface substrate with a microfluidic (MF) chip made of transparent polydimethylsiloxane (PDMS), a metasurface FL sensor was formed (Figure 1d), which is illuminated by the green LED light. The outlook of the experimental setup, including the MF and optical elements, is shown in Figure 1e.

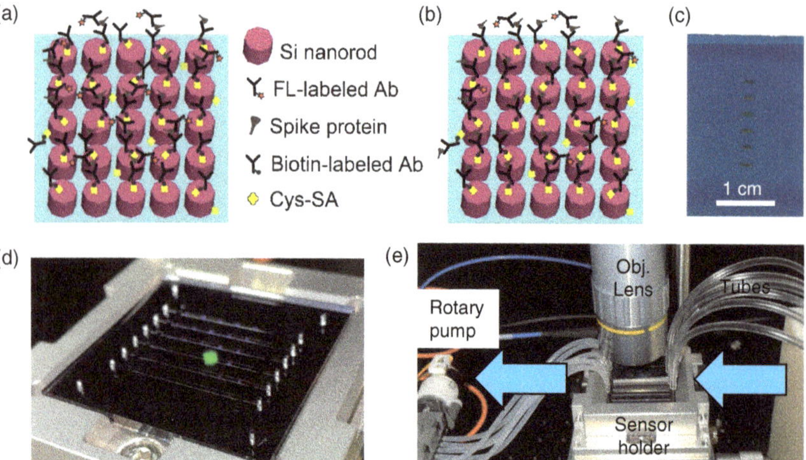

Figure 1. Schematics and overview of this study. Illustrations for detection of (**a**) glycoproteins and (**b**) antibodies (Abs) on the all-dielectric metasufaces of the Si nanorod array, which is located on the SiO_2 layer (pale blue). Si nanorods, labeled Abs, spike protein, and Cys-SA are listed. (**c**) Photograph of an all-dielectric metasurface substrate. Six metasurface areas are vertically arrayed. The white scale bar indicates 1 cm. (**d**) Metasurface sensor chip, composed of a self-absorbed pair of an on-top PDMS microfluidic chip with six channels and a metasurface substrate of 45×45 mm^2. The chip is set in a holder. (**e**) Actual view of experimental setup around a metasurface sensor chip in the holder. The metasurface sensor chip is connected through tubes. The arrows indicate the direction of liquid flow. The objective lens facilitates access from the top to conduct FL measurement on the metasurface sensor.

2. Materials and Methods

2.1. Biosamples

As a testbed for the proof-of-concept experiment, a COVID-19 glycoprotein peptide (ab273063, Abcam, Cambridge, UK) and the corresponding Ab (ab272504, Abcam) were chosen. The glycoprotein peptide with the C terminal was 20 amino acid peptides within the last 50 amino acids of the COVID-19 spike glycoprotein, allowing the rabbit-polyclonal Abs to bind to the spike glycoprotein of COVID-19. The molecular weight (MW) was measured in an electrophoresis setup, and it was confirmed that the Abs had almost the same MW to IgG Abs (Supplementary Materials: Figure S1).

The Abs were originally unconjugated. We labelled the Abs for the sandwich assay, as depicted in Figure 1a,b. 50 μg Ab was used for biotin labeling using a commercial kit (LK03, Dojindo Molecular Technologies, Kumamoto, Japan). Another 50 μg Ab was labeled with HyLite Fluor 555 (HL555) fluorescence molecules using a kit (LK14, Dojindo Molecular Technologies). HL555 has a light absorption peak at 555 nm and emits FL at the wavelength range of 560–660 nm. The HL555-labeling ratio was evaluated by measuring the transmittance of HL555-labeled Abs, which was found to be approximately 7 HL555-molecules/Ab (Supplementary Materials: Figure S2). After labeling, the concentration of HL555-Ab was 0.5 mg/mL in phosphate-buffered saline (PBS, 164-25511, Fujifilm Wako Pure Chemical, Osaka, Japan) of pH 7.4. Biotin- and HL555-labeled Abs were preserved at 5 °C for use in sandwich assay detection. Dilution was conducted each day. The binding of the Abs on Si-nanorod metasurfaces was efficiently conducted using binding molecules of Cys-streptavidin (Cys-SA, PRO1005, ClickBiosystems, Richardson, TX, USA). The detection configurations using the labeled Abs and Cys-SA are shown in Figure 1a,b.

2.2. Metasurface Design and Fabrication

The all-dielectric metasurfaces employed in this study (Figure 2a) were conceived by introducing nanostructures in a thin Si waveguide with 200 nm thickness on a SiO_2 layer. The silicon-on-insulator (SOI) waveguide is optically isolated, thereby enabling us to study the light-confined modes, which are the source of prominent electromagnetic resonances in all-dielectric metasurfaces [12]. A concrete nanostructure was designed on the basis of the numerical simulations for optical properties using a rigorous coupled-wave analysis [13] and a scattering-matrix algorithm [14]. Figure 2b shows the reflectance spectrum at normal incidence, which was computed for a metasurface of periodicity 300 nm and Si-nanorod-diameter 220 nm; the height of Si nanorod was set to 200 nm, which is equal to the thickness of the SOI layer. Optical resonances appear as reflectance peaks or dips, which were tuned to the HL555-FL-emitting wavelengths of 560–660 nm by adjusting the structural parameters in an empirical manner and through confirmation by numerical simulations. Resonant electromagnetic fields were visualized in the numerical simulation. The resonant fields at 765.1 and 576.5 nm are presented in Figure 2c–f, respectively.

The fabrication of the all-dielectric metasurfaces was conducted via a nanolithography process using electron-beam lithography and dry etching on the SOI wafers. The procedure has already been specified in previous publications [9–11]. In this study, the size of metasurface substrates was set to 45 × 45 mm^2 and each substrate had six metasurface areas, as shown in Figure 1c, corresponding to the six MF channels. We note that the metasurface substrates were reusable after washing with an acid solution [15].

2.3. MF Protocols

As shown in Figure 1e, sample liquid arrived through tubes, flowed on the metasurface, and exited through a small rotary pump (RP-6R01S-5A-DC3V, Takasago Fluidic Systems, Nagoya, Japan). The MF system is important to ensure the quantitative control of flow rate of liquids in the metasurface sensors. The flow rate was set to 7–80 μL/min, in accordance with the liquid species noted in Section 2.1. The small rotary pump has six channels, and the flow-rate variation between the channels was approximately 5% at designated rates. The PDMS MF chip has six flow channels and forms flow paths, including metasurface areas, when attached to the metasurface substrate in a self-absorbed manner. Each channel has inlet and outlet holes at both ends. The height of the MF flow paths was designed to be 30 μm, and the thickness of the PDMS chip was 2 mm. Liquid samples accessed the MF flow paths through a stainless pin connected to the tubes outside the metasurface sensor (Figure 1e).

In this study, we used the two configurations shown in Figure 1a,b. The target was the glycoprotein peptide in the former configuration, and it was assumed to have diverse target concentrations; indeed, it was diluted to a level less than pg/mL. In the latter configuration, the glycoprotein Ab was the target and varied concentrations, whereas the capture Ab labeled with biotin and the glycoprotein were densely immobilized, flowing at certain concentrations at μg/mL level. Generally, the high-concentration proteins, including Abs, only need a short time (typically, 5 min) of flow at 10–12 μL/min, whereas the low-concentration targets require a low flow rate of 7–10 μL/min for 20 min. Preflow to fill the MF channels with PBS was performed at 70–80 μL/min.

On the basis of these conditions, we conducted the MF protocol for glycoprotein detection on each metasurface as follows: PBS preflow at 400 ± 30 μL, Cys-SA flow 124 ± 4 μL, PBS rinse 273 ± 13 μL, biotin-labeled Ab 142 ± 6 μL, PBS rinse 188 ± 4 μL, glycoprotein peptide 146 ± 6 μL, PBS rinse 221 ± 9 μL, HL555-labeled Ab 134 ± 6 μL, PBS rinse 160 ± 5 μL, and finally a PBS-T (163-24361, Fujifilm Wako Pure Chemical, Osaka, Japan) rinse 275 ± 7 μL. The concentrations of Cys-SA, biotin-labeled Ab, and HL555-labeled Ab were set to 2 μg/mL, 2.5 μg/mL, and 100 ng/mL, respectively. Cys-SA and the labeled Abs were diluted with PBS and diluent NS buffer (ab193972, Abcam), respectively. The target glycoprotein was diluted with PBS.

We refer to the background image that was taken just after the first PBS rinse, and FL images were acquired after the final rinse. The net FL intensities were evaluated by subtracting the background level from the recorded FL images. We mention here that the MF flows required considerable time, approximately 2 h; however, this is shorter than the time needed for conventional ELISA. The MF runtime can be shortened by reducing the rinse time; for example, the PBS rinse just before the PBS-T rinse can be omitted. Total MF time is estimated to be finished within 80 min, and further reduction can be realized to confirm the lowest necessary amount of Cys-SA and labeled Abs in this assay.

Precoating of capture Abs is valid to realize a short-time detection for spike proteins for practical purposes. The biotin-labeled Abs were immobilized in advance and preserved at 5 °C for approximately 20 h. Afterwards, glycoprotein detection was conducted within 1 h. The MF protocol was similar to that noted above.

Ab detection was conducted as follows: PBS preflow 427 ± 7 μL, Cys-SA 126 ± 2 μL, PBS rinse 267 ± 13 μL, biotin-Ab 130 ± 25 μL, PBS rinse 182 ± 5 μL, glycoprotein peptide 146 ± 4 μL, PBS rinse 217 ± 3 μL, HL555-Ab 138 ± 5 μL, PBS rinse 169 ± 3 μL, and a final PBS-T rinse 284 ± 6 μL. In this assay, the target was HL555-labeled Ab flowing at 100, 25, 6.25, 1,56, 0.39, and 0 ng/mL, where the zero concentration means negative control. The concentration of the glycoprotein was fixed at 1 μg/mL, because it played the role of the capture molecule. The concentrations of Cys-SA and the biotin-labeled Ab were the same as those in the glycoprotein detection.

2.4. Optical Measurement

FL images of 16-bit depth were acquired using a MF setup equipped with an uncooled CCD camera (Infinity-3S, Teledyne-Lumenera, Ottawa, ON, Canada) which was attached with a 10× objective lens of numerical aperture (NA) 0.28 (M Plan Apo, Mitsutoyo, Kawasaki, Japan). Excitation light came from a green LED (M530F2, Thorlabs, Newton, NJ, USA) and was spectrally reshaped through bandpass filters transmitting a band of 516–538 nm. The green spot (Figure 1d) was introduced with the LED and the objective lens. The FL emitted on the metasurfaces was collected by the objective lens, passed through FL bandpass filters transmitting light of 567–616 nm wavelengths, and was detected by the CCD camera. In the FL measurement, the background level was not as low because the uncooled CCD camera was operated with a gain mode. A typical FL measurement was conducted under conditions of 3 s exposure and 10× gain. Thus, we note that the limit of detection (LOD) was limited by the FL-detection system and not the metasurface sensors.

To examine LOD of the metasurface sensors themselves, an extremely low background FL measurement was conducted in an upright confocal FL microscope (Stellaris 5, Leica Microsystems, Wetzlar, Germany). Low-background measurement was conducted with operating detectors in the photon counting mode. To cover a metasurface area of mm^2 dimension, a $5\times$ objective lens of NA 0.15 was used. The excitation wavelength for HL555 was set to 521 nm, the lateral resolution in the confocal microscopy was $0.51 \times 521/NA = 1771.4$ nm in air, and the detection wavelength was set to a range of 560–620 nm.

3. Results

3.1. Optical Properties of the Metasurfaces

Figure 2 shows the structural and optical properties of the all-dielectric metasurface. A typical top-view scanning electron microscopy (SEM) image of the metasurface is shown in Figure 2a (a wide view with a scale bar of 5 µm (white bold) and a magnified view (inset) with a scale bar of 500 nm (white)). The periodicity was designed to be 300 nm, and the diameter of Si nanorod was set to 220 nm. The SEM image confirmed that the periodicity is almost exactly 300 nm and the diameter is 220 ± 3 nm.

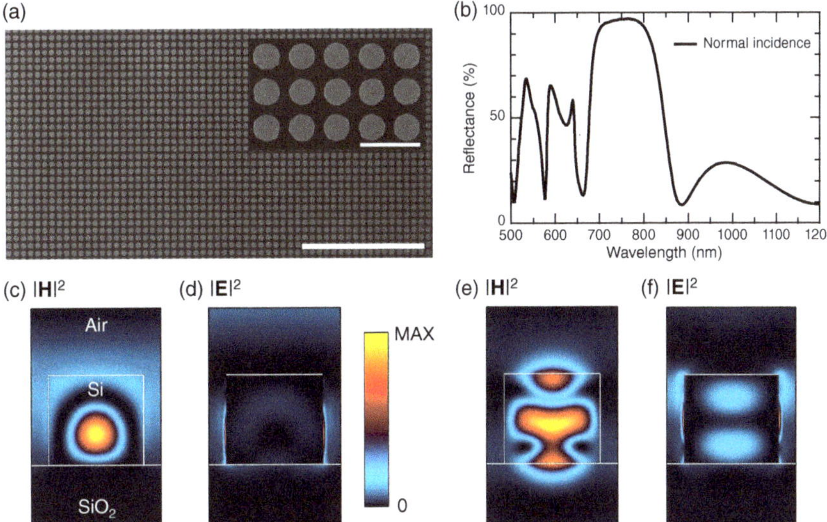

Figure 2. Structure, optical properties, and resonances of all-dielectric metasurfaces. (**a**) SEM image in wide view. The white bold scale bar indicates 5 µm. The inset is shown in a magnified view with a white scale bar of 500 nm as well. The periodicity of the metasurface is 300 nm and diameter of the Si nanorod is 220 ± 3 nm. (**b**) Reflectance spectrum at normal incidence, which was numerically simulated. (**c**,**d**) Respective resonant magnetic and electrical field intensities at 765.1 nm. The vertical section view through the center of the Si nanorod is shown together with the color bar. (**e**,**f**) Respective resonant magnetic and electrical field intensities at 576.5 nm, shown in a similar manner to (**c**,**d**).

The optical features of the metasurface can be seen in the reflectance spectrum in Figure 2b. In accordance with the short period length of 300 nm and the 200 nm height of the Si nanorods, the lowest resonant mode in photon energy, which is the longest wavelength mode, appears as a high-reflectance band at 700–800 nm. The lowest mode is ascribed to the magnetic dipole mode because the resonant magnetic field has a single node inside the Si nanorod, as shown in Figure 2c. The electric field is strongly localized at the sidewall of the Si nanorod, as shown in Figure 2d; the distribution is probably suitable for enhancing the electric-dipole transitions, resulting in FL molecules. The magnetic and electric field intensities reach 58.8 and 58.1, respectively, when setting the incident field intensity to 1. Thus, prominent field enhancement is observed on the resonance.

The HL555 molecules have their main FL band at 570 nm. Therefore, it is highly preferable that the metasurface have a resonant magnetic mode around this wavelength. As seen in the reflectance spectrum in Figure 2b, a reflectance dip appears at 576.5 nm. The resonant magnetic and electrical field intensities are shown in Figure 2e,f, respectively. The magnetic field distribution indicates a higher magnetic mode, and the electrical field is strongly localized at the outermost surface of the Si nanorod. The maximum intensities of magnetic and electric fields, when the incident intensities are 1, are 69.8 and 57.5, respectively. This resonance was tuned to enhance the FL emission of the HL555 molecules. It is noteworthy that the total observed enhancement of the FL process was determined by three key factors: excitation efficiency, inner quantum yield in the FL molecule, and FL emission efficiency [16–18]. The enhancement of FL emission efficiency is often called the Purcell effect, which describes the expedited rate of electronic transition in resonant electromagnetic fields [19]. The observed enhancement in FL intensity originates from the abovementioned three factors, and the total optimization is essential [20].

3.2. Glycoprotein Peptide Detection

Figure 3 shows the spike glycoprotein peptide detection in the configuration illustrated in Figure 1a. The target concentrations were changed from 100 to 0 ng/mL. Figure 3a shows the detected FL images at the nonzero target concentrations, represented with raw color; the brightness was increased by 40% and the contrast was decreased by −40% for better visibility. The FL-emitting areas correspond to the all-dielectric metasurfaces inside the MF channels. Note that although FL-labeled Abs were flowed inside the MF channels, the definite FL was observed only on the metasurfaces owing to their high capability of enhancing FL intensity [18].

The FL intensities extracted from the images in Figure 3a are plotted in Figure 3b,c; the former is a semi-logarithmic representation and the latter is a linear one. The red closed circles denote the FL intensities, which are shown with error bars. The data point at 0 ng/mL can be present in the linear presentation, which is a negative control used to identify the zero level in the measurement. The black curves are fitted detection curves, and are identical to each other in Figure 3b,c, although their appearance is different in the two representations. The detection curve is described by the Hill equation [21]:

$$y = y_0 + (S - y_0)\frac{x^n}{x^n + K_D^n} \quad (1)$$

where y denotes the FL intensity, y_0 is the background level without any target, S is the saturation signal intensity, which was regarded as a proportional constant in fitting, x is the concentration of target, n is the degree of cooperative reaction, and K_D is the dissociation constant [22,23]. The parameter n suggests an anti-cooperative binding reaction for $n < 1$ and a cooperative reaction for $n \geq 1$ [24]. In fitting the data in Figure 3b, the baseline y_0 was assumed to be 0, while the variables were n, K_D, and S. By fitting using commercial graphic software [25], we found that $n = 0.29$, $K_D = 30.5$ ng/mL, and $S = 10277$. These results suggest that the binding reaction is anti-cooperative, which is often observed between antigens and Abs [9,11]. The Hill equation is mathematically equivalent to the so-called four-parameter logistic equation [9]. From the Hill curve and 3σ level of the dotted line in Figure 3b, where σ is the standard deviation, the LOD was estimated to be 0.8 pg/mL, which is in principle reachable in a very low background measurement.

As described in Section 2.4, low-background FL measurement is generally needed to identify the LOD of the metasurface sensor. Figure 3d shows a set of confocal FL images at the target concentrations of 400, 16, and 0.64 pg/mL. The brightness and contrast of the FL images are set in common. The confocal images are originally grayscale and shown in pseudocolor. The rectangular areas (colored) of 2.1×0.7 mm^2 are the metasurface. The MF-flow direction was from right to left; as a result, the right-hand side is relatively bright in all the images, indicating that the immobilization of the biomolecules took place from the right-hand side (or the inlet side of the MF channels). Thus, the confocal FL images provide further insight due to the higher lateral resolution.

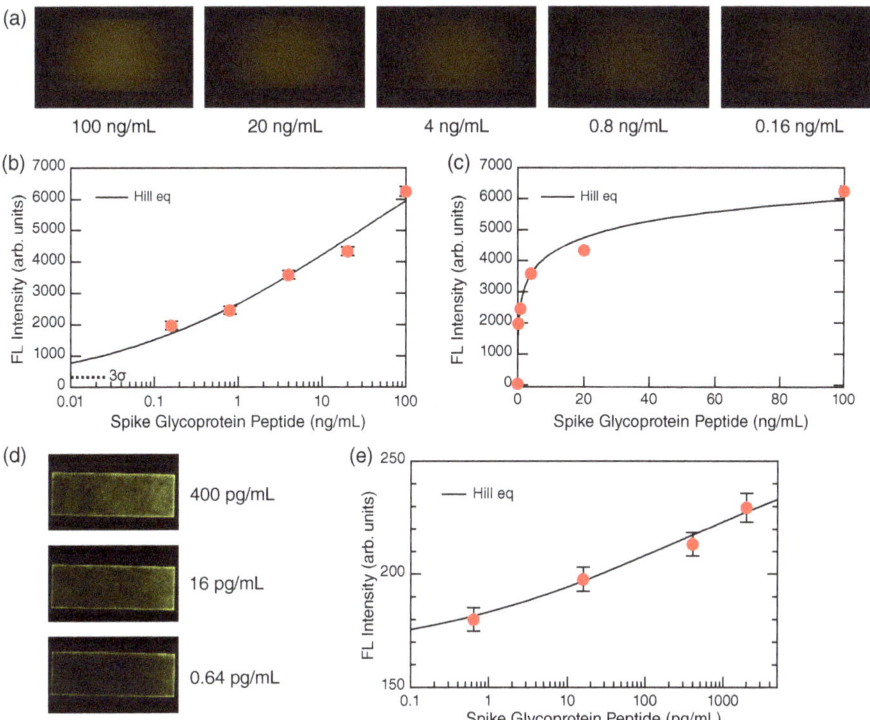

Figure 3. Glycoprotein detection. (**a**) FL images (raw color) of the target concentration from 100 to 0.16 ng/mL. (**b**) Detection data plot in the semi-logarithmic representation. Red dots indicate the FL intensities extracted from the images in (**a**). Error bars are shown together. The solid curve is a fitted curve following the Hill equation (Equation (1)). The level of 3σ from zero is indicated by the dotted line. (**c**) The detection data plot is shown in a linear representation with data in common with (**b**), except for the data point at 0 ng/mL. (**d**) Confocal FL images at low concentrations from 400 to 0.64 pg/mL are shown. The images were originally grayscale, shown by pseudocolor. The colored rectangular areas of 2.1×0.7 mm^2 correspond to the metasurfaces. The MF flow was conducted from right to left. (**e**) Detection curve obtained from the confocal images. Note that the abscissa is in units of pg/mL. A curve fitted by the Hill equation (Equation (1)) is shown with the black curve.

The FL intensities in the confocal images were evaluated with settings for analysis at a common analyzing area (Supplementary Materials: Figure S3). The intensities are plotted in Figure 3e as red closed circles with error bars; a fitted curve using the Hill equation (Equation (1)) is shown with a black curve. In the fitting, the parameters n, K_D, and S were found to be 0.26, 273.6 pg/mL, and 265.7, respectively. The concentration range in Figure 3e is limited to a low range; therefore, K_D and S tend to depend on the concentration range when it is narrow, indeed shifting to smaller values than those found in Figure 3b,c. However, n, representing the curvature of the fitted Hill curve, is quite close to the parameter n in Figure 3b,c, which consistently suggests that the binding reaction of FL-labeled Abs with the target proteins is anti-cooperative. The confocal FL imaging shows that the metasurface sensor is capable of detecting the target even at 0.64 pg/mL, and indicates that the intrinsic LOD is located in the sub-pg/mL range, which is consistent with the estimation using the 3σ line in Figure 3b. In terms of the dynamic range, the metasurface FL sensors were revealed to have a dynamic range more than five orders of the target concentrations, because the FL signals were detected from 100 ng/mL to 0.64 pg/mL in the scaled manner. Although the measured data are a different set from the above, we show confocal FL images at 0.49 and 0 pg/mL (Supplementary Materials: Figure S4). The image at 0 pg/mL indicates the signal level of the negative control. It is shown in the supplementary materials that the FL-signal level at 0.49 pg/mL is located at more than 1σ from that of the negative control while it is within 3σ; therefore, the LOD is located above 0.49 pg/mL, and near 0.64 pg/mL.

Figure 4 shows a series of results on precoating the capture Abs in advance. If the precoating is successful, it becomes possible to substantially reduce detection time in practical situations. Two configurations in the precoating and preservation are illustrated in Figure 4a,b, which have elements in common with Figure 1a. In the former, the capture Abs with the biotin label were preserved in PBS, and in the latter, the capture Ab was dried in air and preserved. The two-type precoated metasurfaces were kept at 5 °C for approximately 20 h.

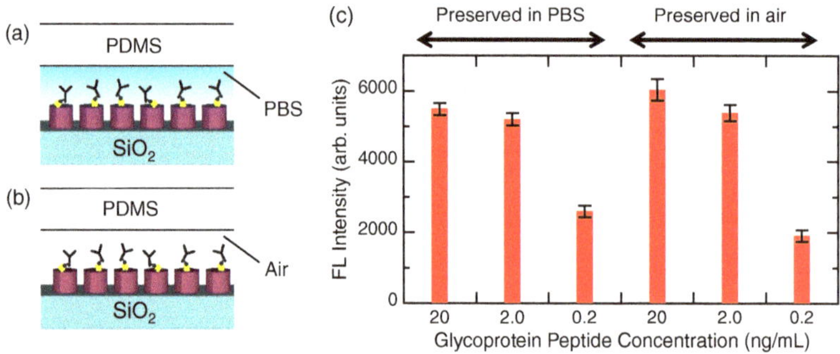

Figure 4. Test for precoating of the capture Abs. (**a**) Precoated Abs were kept in PBS for ~20 h. (**b**) Precoated Ab was kept in air and dried for ~20 h. (**c**) After a short runtime of the target glycoprotein peptide and the FL-labeled detection Ab, the detected FL intensities are shown using red bars with error bars in the two cases (**a**,**b**).

After preservation, PBS was flowed at 40–45 µL/min for 6 min in all the channels. Subsequently, the glycoprotein, PBS, the HL555-Ab, and PBS-T were flowed within 1 h, in a manner similar to that used for glycoprotein detection in Figure 3. After the flow, the FL images were measured. Figure 4c shows the FL intensities. The target was detected in both cases, indicating that precoating is possible in the assay using the metasurface sensors. We found that preservation in air led to inhomogeneous FL spots; therefore, the recovery process of the dried Ab after preservation should be considered further.

3.3. Ab Detection

Figure 5 shows a set of experimental results from the Ab detection illustrated in Figure 1b. The glycoprotein was densely immobilized on the metasurface sensor, and the Abs at different concentrations were detected. In this proof-of-concept experiment, the target Abs were labeled with HL555, and direct detection of the Abs was conducted on this basis.

Three FL images at different Ab concentrations from 100 to 6.25 ng/mL are shown in Figure 5a; the representation setting is similar to that in Figure 3c. The detection profile is shown in Figure 5b. The FL intensities (red closed circles) were evaluated from the measured FL images, shown with error bars. Obviously, the profile is almost linear, with a variance of $R^2 = 0.99$. The inset magnifies a low-concentration range and shows that the linear response holds to 0 ng/mL. Thus, the linear response is a good feature to provide a quantitative assay for the Abs.

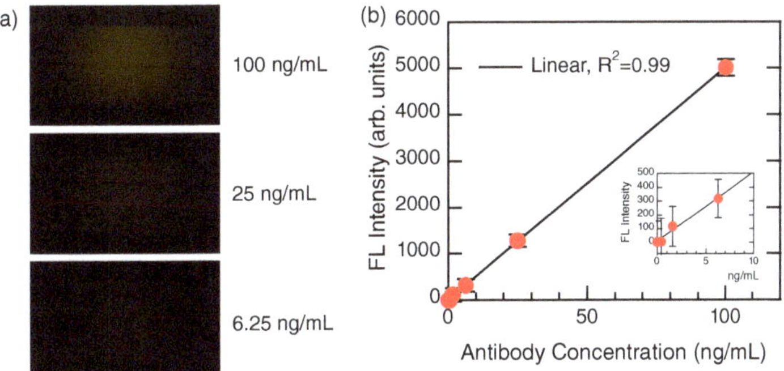

Figure 5. Detection of Abs, as illustrated in Figure 1b. (**a**) FL images (raw color) at different Ab concentrations from 100 to 6.25 ng/mL (or 686 to 41.2 pM). (**b**) Linear plot of the FL intensity for the Ab concentrations. The measured data are shown with red closed circles with error bars. The black line is linearly fitted. The inset magnifies a low-concentration range.

4. Discussion

4.1. Glycoprotein Peptide Detection

It was demonstrated that the metasurface FL sensor is capable of a very low target concentration of 0.64 pg/mL. We employed confocal FL microscopy to perform extremely low background measurement, observing that even the low concentrations could be detected in a scaled manner, as shown in Figure 3e. This suggests that the LOD of the metasurface FL sensor itself exists at a lower concentration than the measured concentration of 0.64 pg/mL.

In a practical setup using an uncooled CCD camera, the target glycoprotein peptide was detected down to approximately 0.1 ng/mL. The detection range was limited by the performance of the CCD camera. However, this detection range is better than that of conventional ELISA, which shows an LOD of 0.5 ng/mL [26]; the dynamic range was estimated to be 1–64 ng/mL from open data. In terms of detection runtime, the present method is substantially faster than conventional ELISAs, which usually take at least several hours.

Table 1 lists detection performance on the several platforms addressed in this article. Here, we define dynamic range as a range scaled by a rational equation and distinguish signals between one-order different target concentrations in the 1σ criterion. Runtime means all the time that is needed for a test, and is inseparable.

Table 1. COVID-19 detection for proteins and antibody. ICH denotes immunochromatography-based disposable test kits. NA denotes not available. Dynamic range defined in the text is estimated from the published data in the references. Metasurface sensor refers to this study. Runtime includes all the preparation time that is inseparable in a test.

Target	Method	Dynamic Range	Runtime
Spike proteins	ICH	NA	20–30 min
Spike glycoproteins	ELISA [26]	1–64 ng/mL	4–6 h
Spike glycoproteins	Metasurface sensor	0.64–100,000 pg/mL	<1 h
Glycoprotein Ab (IgG-type)	Metasurface sensor	1.56–100 ng/mL (or 10.7–686 pM)	30 min
IgG	Resonant shift [27]	5–100 nM	70 min
IgG	Resonant shift [28]	1–10 nM	overnight

4.2. Ab Detection

In Ab detection, the IgG-type Ab was detected even at 1.56 ng/mL, which is 10.7 pM, in a linear manner. Table 1 lists the Ab detection results. IgG-type COVID-19 Ab was detected at a range of 5–100 nanomolars (nM) in a previous report using gold nanoparticles [27], where the linear dynamic range was limited to a range from 100 nM to approximately 10 nM, which was claimed to meet the detection of neutralizing Abs. Thus, the present metasurface sensors exhibit approximately 100-fold better detection performance, thereby being sufficient for qualifying the neutralizing Abs.

Next, we discuss the robustness of the all-dielectric metasurface sensors for impeding biomolecules. It has been reported that Ab detection in metasurface FL sensors is robust when human serum is included in the target buffer [9]. As is widely known, human serums contain abundant biomolecules such as albumin, lipin, IgG, and more; for example, IgG is typically included at 87–170 mg/mL. In the previous publication [9], it was demonstrated that the target IgG at 10 ng/mL can be detected without substantial loss. Thus, even when the impeding IgG is more than 100,000-fold more dense than the target IgG, all-dielectric metasurface sensors can detect the target. Although we focus here on demonstrating the detection of glycoproteins and Abs, we believe that the metasurface sensors retain this robustness.

For optical detection of proteins such as Abs, resonant wavelength-shift measurements are often reported. Regarding COVID-19 IgG, a very low LOD of 30 aM has been reported [28]; such low LODs have sometimes been claimed for other biomolecules as well [29]. However, the amount of wavelength shift for the COVID-19 IgG is < 10 nm for eight-order concentration changes. As a result, similar concentrations are often indistinguishable; for example, the experimental data indicate that the method cannot discriminate 1 nM from 0.1 nM or 0.1 nM from 0.01 nM [28]. Thus, the technique has a disadvantage in quantifying the target IgG concentrations even at sub-nM ranges as compared with the present metasurface FL sensors. Resonant-shift assays stemming based on surface plasmon resonance [30] have been studied for more than thirty years and have yet to be established as a high-sensitivity method, likely for this very reason.

4.3. Further Designs for Metasurface FL Sensors

There are many reports on non-empirical designs for metasurfaces [31–35]. Nevertheless, trials for FL-enhancing metasurfaces have hardly been reported. This is partially because FL enhancement involves excited electronic states in molecules, and simulations for the metasurface design cannot handle the excited states. To implement simulations including excited-state dynamics, it is necessary to incorporate light–matter interaction. Thus, designs for metasurface FL sensors are generally difficult. However, in Section 3.1 we have described features in electromagnetic resonance that are suitable for efficient FL enhancement; therefore, we consider that clues to finding other structures suitable for FL sensing are provided by the features in the resonant modes.

5. Conclusions

In this study, we performed a series of proof-of-concept experiments to detect COVID-19 spike glycoproteins and Abs in one platform, i.e., an all-dielectric metasurface FL sensor. In a practical setup, the glycoprotein was detected more efficiently than when using conventional ELISA. In addition, the LOD of the metasurface sensors was tested using confocal microscopy; the LOD was found to be <0.64 pg/mL, and the dynamic range was more than five orders of the target concentrations. Thus, the metasurface FL sensors have extremely high-sensitivity. Moreover, the Ab detection was linear down to $\sim$10 pM, which, to our knowledge, is highly reliable in terms of quantification ability. The Ab detection range is sufficient for testing the neutralizing Abs. As discussed in Section 4.2, all-dielectric metasurface sensors are robust for impeding biomolecules. Overall, the present method is highly promising for straightforward extension to practical situations.

Supplementary Materials: The following are available online at https://www.mdpi.com/article/10.3390/bios12110981/s1, Figure S1: molecular weight analysis, Figure S2: transmission spectra of HL555-labeled Ab, Figure S3: FL-image analysis with setting an analyzing box, Figure S4: confocal FL images including a negative control. References [36–38] are cited in the supplementary materials.

Author Contributions: Conceptualization, supervision of the whole study, construction of the MF system, measurement using confocal FL microscopy, data analysis, numerical implementation, nanofabrication, and draft, M.I. Biosample preparation, experiment using the MF system and the electrophoresis setup, and initial data analysis, W.T. All authors have read and agreed to the published version of the manuscript.

Funding: M.I. acknowledges financial supports by M-Cube project and the support system for curiosity-driven research in NIMS.

Data Availability Statement: Data in this article are available from the corresponding author upon reasonable request.

Acknowledgments: We appreciate Naoki Ikeda for technical support in the nanofabrication of the metasurfaces and Xianglan Li for technical support in the electrophoresis experiment. A part of this study was conducted at Namiki Foundry and NanoBio Laboratory in NIMS. Numerical implementation was executed at SX-AOBA in Cyber Science Center, Tohoku University.

Conflicts of Interest: The authors declare no conflict of interest.

References

1. Saiki, R.K.; Scharf, S.; Faloona, F.; Mullis, K.B.; Horn, G.T.; Erlich, H.A.; Arnheim, N. Enzymatic Amplification of β-Globin Genomic Sequences and Restriction Site Analysis for Diagnosis of Sickle Cell Anemia. *Science* **1985**, *230*, 1350–1354. [CrossRef] [PubMed]
2. Udugama, B.; Kadhiresan, P.; Kozlowski, H.N.; Malekjahani, A.; Osborne, M.; Li, V.Y.C.; Chen, H.; Mubareka, S.; Gubbay, J.B.; Chan, W.C.W. Diagnosing COVID-19: The Disease and Tools for Detection. *ACS Nano* **2020**, *14*, 3822–3835. [CrossRef] [PubMed]
3. Ji, T.; Liu, Z.; Wang, G.; Guo, X.; Khan, S.A.; Lai, C.; Chen, H.; Huang, S.; Xia, S.; Chen, B.; et al. Detection of COVID-19: A review of the current literature and future perspectives. *Biosens. Bioelectron.* **2020**, *166*, 112455. [CrossRef] [PubMed]
4. Zuo, Y.Y.; Uspal, W.E.; Wei, T. Airborne Transmission of COVID-19: Aerosol Dispersion, Lung Deposition, and Virus-Receptor Interactions. *ACS Nano* **2020**, *14*, 16502–16524. [CrossRef] [PubMed]
5. Noviana, E.; Ozer, T.; Carrell, C.S.; Link, J.S.; McMahon, C.; Jang, I.; Henry, C.S. Microfluidic Paper-Based Analytical Devices: From Design to Applications. *Chem. Rev.* **2021**, *121*, 11835–11885. [CrossRef]
6. Lee, C.Y.; Degani, I.; Cheong, J.; Weissleder, R.; Lee, J.H.; Cheon, J.; Lee, H. Development of Integrated Systems for On-Site Infection Detection. *Acc. Chem. Res.* **2021**, *54*, 3991–4000. [CrossRef]
7. Pirzada, M.; Altintas, Z. Nanomaterials for virus sensing and tracking. *Chem. Soc. Rev.* **2022**, *51*, 5805–5841. [CrossRef]
8. Alafeef, M.; Pan, D. Diagnostic Approaches For COVID-19: Lessons Learned and the Path Forward. *ACS Nano* **2022**, *16*, 11545–11576. [CrossRef]
9. Iwanaga, M. All-Dielectric Metasurface Fluorescence Biosensors for High-Sensitivity Antibody/Antigen Detection. *ACS Nano* **2020**, *14*, 17458–17467. [CrossRef]
10. Iwanaga, M. High-Sensitivity High-Throughput Detection of Nucleic-Acid Targets on Metasurface Fluorescence Biosensors. *Biosensors* **2021**, *11*, 33. [CrossRef]
11. Iwanaga, M. Highly sensitive wide-range target fluorescence biosensors of high-emittance metasurfaces. *Biosens. Bioelectron.* **2021**, *190*, 113423. [CrossRef]
12. Kuznetsov, A.I.; Miroshnichenko, A.E.; Brongersma, M.L.; Kivshar, Y.S.; Luk'yanchuk, B. Optically resonant dielectric nanostructures. *Science* **2017**, *354*, aag2472. [CrossRef]
13. Li, L. New formulation of the Fourier modal method for crossed surface-relief gratings. *J. Opt. Soc. Am. A* **1997**, *14*, 2758–2767. [CrossRef]
14. Li, L. Formulation and comparison of two recursive matrix algorithm for modeling layered diffraction gratings. *J. Opt. Soc. Am. A* **1996**, *13*, 1024–1035. [CrossRef]
15. Choi, B.; Iwanaga, M.; Ochiai, T.; Miyazaki, H.T.; Sugimoto, Y.; Sakoda, K. Subnanomolar fluorescent-molecule sensing by guided resonances on nanoimprinted silicon-on-insulator substrates. *Appl. Phys. Lett.* **2014**, *105*, 201106. [CrossRef]
16. Choi, B.; Iwanaga, M.; Miyazaki, H.T.; Sugimoto, Y.; Ohtake, A.; Sakoda, K. Overcoming metal-induced fluorescence quenching on plasmo-photonic metasurfaces coated by a self-assembled monolayer. *Chem. Commun.* **2015**, *51*, 11470–11473. [CrossRef]
17. Iwanaga, M.; Choi, B.; Miyazaki, H.T.; Sugimoto, Y. The artificial control of enhanced optical processes in fluorescent molecules on high-emittance metasurfaces. *Nanoscale* **2016**, *8*, 11099–11107. [CrossRef]
18. Iwanaga, M. All-Dielectric Metasurfaces with High-Fluorescence-Enhancing Capability. *Appl. Sci.* **2018**, *8*, 1328. [CrossRef]
19. Saleh, B.E.A.; Teich, M.C. *Fundamentals of Photonics*, 2nd ed.; Wiley-Interscience: Oxford, UK, 2007; pp. 515–516.
20. Iwanaga, M. *Plasmonic Resonators: Fundamentals, Advances, and Applications;* Pan Stanford Publishing: Singapore, 2016; Chapter 5.
21. Hill, A.V. The possible effects of the aggregation of the molecules of haemoglobin on its dissociation curves. *J. Physiol.* **1910**, *40*, iv–vii.

22. Neubig, R.R.; Spedding, M.; Kenakin, T.; Christopoulos, A. International Union of Pharmacology Committee on Receptor Nomenclature and Drug Classification. XXXVIII. Update on Terms and Symbols in Quantitative Pharmacology. *Pharmacol. Rev.* **2003**, *55*, 597–606. [CrossRef]
23. Gesztelyi, R.; Zsuga, J.; Kemeny-Beke, A.; Varga, B.; Juhasz, B.; Tosaki, A. The Hill equation and the origin of quantitative pharmacology. *Arch. Hist. Exact Sci.* **2012**, *66*, 427–438. [CrossRef]
24. Irrera, A.; Leonardi, A.A.; Di Franco, C.; Lo Faro, M.J.; Palazzo, G.; D'Andrea, C.; Manoli, K.; Franzò, G.; Musumeci, P.; Fazio, B.; et al. New Generation of Ultrasensitive Label-Free Optical Si Nanowire-Based Biosensors. *ACS Photonics* **2018**, *5*, 471–479. [CrossRef]
25. Igor Pro. Available online: https://www.wavemetrics.com/products/igorpro (accessed on 28 October 2022).
26. Antibody Information. Available online: https://www.abcam.com/sars-cov-2-spike-glycoprotein-antibody-coronavirus-ab272504.html (accessed on 20 September 2022).
27. Lew, T.T.S.; Aung, K.M.M.; Ow, S.Y.; Amrun, S.N.; Sutarlie, L.; Ng, L.F.P.; Su, X. Epitope-Functionalized Gold Nanoparticles for Rapid and Selective Detection of SARS-CoV-2 IgG Antibodies. *ACS Nano* **2021**, *15*, 12286–12297. [CrossRef] [PubMed]
28. Masterson, A.N.; Sardar, R. Selective Detection and Ultrasensitive Quantification of SARS-CoV-2 IgG Antibodies in Clinical Plasma Samples Using Epitope-Modified Nanoplasmonic Biosensing Platforms. *ACS Appl. Mater. Interf.* **2022**, *14*, 26517–26527. [CrossRef] [PubMed]
29. Joshi, G.K.; Deitz-McElyea, S.; Liyanage, T.; Lawrence, K.; Mali, S.; Sardar, R.; Korc, M. Label-Free Nanoplasmonic-Based Short Noncoding RNA Sensing at Attomolar Concentrations Allows for Quantitative and Highly Specific Assay of MicroRNA-10b in Biological Fluids and Circulating Exosomes. *ACS Nano* **2015**, *9*, 11075–11089. [CrossRef]
30. Raether, H. *Surface Plasmons on Smooth and Rough Surfaces and on Gratings*; Springer: Berlin/Heidelberg, Germany, 1988.
31. Ong, J.R.; Chu, H.S.; Chen, V.H.; Zhu, A.Y.; Genevet, P. Freestanding dielectric nanohole array metasurface for mid-infrared wavelength applications. *Opt. Lett.* **2017**, *42*, 2639–2642. [CrossRef]
32. Liu, Z.; Zhu, D.; Rodrigues, S.P.; Lee, K.T.; Cai, W. Generative Model for the Inverse Design of Metasurfaces. *Nano Lett.* **2018**, *18*, 6570–6576. [CrossRef]
33. Iwanaga, M. Non-Empirical Large-Scale Search for Optical Metasurfaces. *Nanomaterials* **2020**, *10*, 1739. [CrossRef]
34. Meem, M.; Banerji, S.; Pies, C.; Oberbiermann, T.; Majumder, A.; Sensale-Rodriguez, B.; Menon, R. Large-area, high-numerical-aperture multi-level diffractive lens via inverse design. *Optica* **2020**, *7*, 252–253. [CrossRef]
35. Hammond, A.M.; Oskooi, A.; Chen, M.; Lin, Z.; Johnson, S.G.; Ralph, S.E. High-performance hybrid time/frequency-domain topology optimization for large-scale photonics inverse design. *Opt. Express* **2022**, *30*, 4467–4491. [CrossRef]
36. Laemmli, U.K. Cleavage of structural proteins during the assembly of the head of bacteriophage T4. *Nature* **1970**, *227*, 680–685. [CrossRef]
37. Manual for FL Labeling Kit. Available online: https://www.dojindo.co.jp/manual/LK14e.pdf (accessed on 21 September 2022).
38. ImageJ Homepage. Available online: https://imagej.nih.gov/ij/ (accessed on 23 September 2022).

 biosensors

MDPI

Review

Strategies for Enhancing the Sensitivity of Electrochemiluminescence Biosensors

Yueyue Huang [1,2], Yuanyuan Yao [1], Yueliang Wang [1], Lifen Chen [1,*], Yanbo Zeng [1], Lei Li [1] and Longhua Guo [1,*]

1 Jiaxing Key Laboratory of Molecular Recognition and Sensing, College of Biological, Chemical Sciences and Engineering, Jiaxing University, Jiaxing 314001, China
2 College of Chemistry and Life Sciences, Zhejiang Normal University, Jinhua 321004, China
* Correspondence: chenlf@zjxu.edu.cn (L.C.); guolh@fzu.edu.cn (L.G.)

Abstract: Electrochemiluminescence (ECL) has received considerable attention as a powerful analytical technique for the sensitive and accurate detection of biological analytes owing to its high sensitivity and selectivity and wide dynamic range. To satisfy the growing demand for ultrasensitive analysis techniques with high efficiency and accuracy in complex real sample matrices, considerable efforts have been dedicated to developing ECL strategies to improve the sensitivity of bioanalysis. As one of the most effective approaches, diverse signal amplification strategies have been integrated with ECL biosensors to achieve desirable analytical performance. This review summarizes the recent advances in ECL biosensing based on various signal amplification strategies, including DNA-assisted amplification strategies, efficient ECL luminophores, surface-enhanced electrochemiluminescence, and ratiometric strategies. Sensitivity-enhancing strategies and bio-related applications are discussed in detail. Moreover, the future trends and challenges of ECL biosensors are discussed.

Keywords: DNA; biosensor; electrochemiluminescence; sensitivity; amplification strategy

Citation: Huang, Y.; Yao, Y.; Wang, Y.; Chen, L.; Zeng, Y.; Li, L.; Guo, L. Strategies for Enhancing the Sensitivity of Electrochemiluminescence Biosensors. *Biosensors* **2022**, *12*, 750. https://doi.org/10.3390/bios12090750

Received: 31 July 2022
Accepted: 8 September 2022
Published: 11 September 2022

1. Introduction

Electrogenerated chemiluminescence (ECL) is the phenomenon resulting from electrogenerated species undergoing an electron-transfer reaction at the electrode surface, resulting in the emission of light [1–3]. ECL has received a large amount of attention due to its advantages of high sensitivity, low background noise, spatial and temporal control, and no required light source [4–7]. The first detailed ECL studies were reported in the 1960s by Hercules and Bard [8,9]. Since then, ECL has gradually become a major area of research, with studies encompassing fundamental studies, reagent development, and analytical applications. Many reviews on the details and on our comprehensive understanding of ECL have been published [10–13]. So far, ECL has been widely applied in various fields, including in food safety, environmental monitoring, and medical diagnosis.

In recent years, ECL biosensors have gradually attracted increasing interest in the field of bioanalysis. Critically, they show great promise for clinical diagnostics and pharmaceutical analysis. Their significant advantages of portability, high sensitivity, and simple operation promote their further development. Moreover, biosensors can provide fast responses at low costs [14–16]. Despite their many merits, the development of biosensors for use in the sensitive and accurate detection of analytes at trace levels with high efficiency and accuracy in complex conditions has represented a critical need in many areas. Specifically, the precision and sensitive measurement of protein biomarkers has great significance and practical value in early diagnosis for disease prediction.

As shown in Scheme 1, biosensing is a process that converts biochemical interactions into output signals for the quantitative determination of target molecules. Double-stranded DNA, single-stranded DNA, antigens, and antibodies are normally employed as recognition elements for biosensor construction. The main signal output modes include single-signal output and multiple-output. Signal amplification is often considered to be one of the most effective strategies for efficient signal transduction and to amplify signal output [17]. Signal amplification-based biosensors ideally possess the features of enhanced sensitivity and selectivity and a wide dynamic range compared to conventional biosensors [18]. Currently, successful signal amplification strategies that are performed in the ECL realm mainly focus on DNA-assisted amplification strategies, improving the efficiency of ECL luminophores and surface-enhanced electrochemiluminescence, ratiometric strategies, and so on. In this review, we have summarized the recently developed and main ECL bioanalysis strategies, with a more detailed emphasis on advanced DNA signal amplification technologies. Finally, the future trends and perspectives of strategies in ECL bioanalysis are briefly outlined.

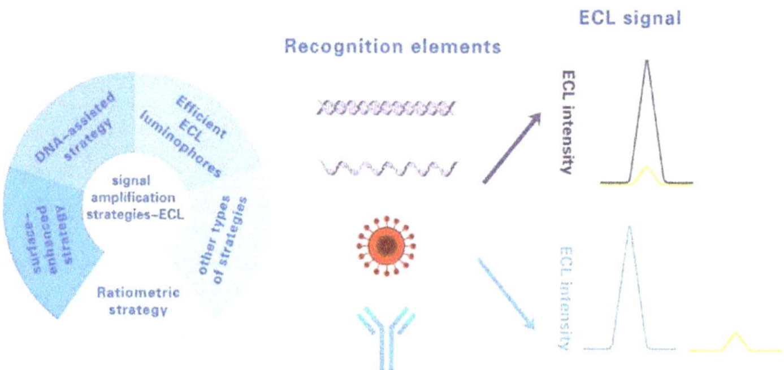

Scheme 1. Overview of signal amplification strategies integrated with ECL biosensors for single-signal or multiple-signal outputs.

2. DNA-Assisted Amplification Strategies

In the past several years, DNA-assisted amplification technologies have received significant attention in biosensing because of their unique structure and properties. Benefiting from the advantages of specific Watson–Crick base pairing and their highly flexible design, DNA molecules can be self-assembled into various DNA structures, such as into DNA dumbbell structures [19], DNA flowers [20], and DNA tetrahedrons [21]. Additionally, signal amplification can be achieved by the governing of the DNA circuits through target triggering using the DNA's programmable operation ability [22]. For example, ECL signal enhancement can be achieved through the target trigger 3D DNA walker moving continuously and automatically along the designed tracks [23,24]. In short, DNA amplification strategies can be classified into two categories: enzyme-assisted amplification and enzyme-free amplification strategies. The former involves enzymes and includes classical polymerase chain reaction (PCR), rolling circle amplification (RCA) or hyperbranched RCA (HRCA), endonuclease- and exonuclease-assisted amplification, and DNAzyme-involved amplification, while hybridization chain reaction (HCR) and DNA walker-based amplification without enzymes are examples of nonenzymatic amplification strategies [25]. Combining these versatile amplification strategies with biosensors can enable remarkable signal enhancements. Several examples that have been reported in recent years are summarized in Table 1, and the details of the signal amplification strategies are discussed below.

Table 1. Examples of representative biosensors based on DNA signal amplification strategies that have been developed in recent years.

Targets	Signal Amplification Strategy	Detection Range	Limit of Detection	Ref.
miRNA	bHCR	0.05–500 fM	0.18 fM	[26]
pyrophosphatase	Cu^{+}-catalyzed azide–alkyne cycloaddition (CuAAC) with high-efficiency hybridization chain reaction (HCR)	0.025–50 mU	8 μU	[27]

Table 1. *Cont.*

Targets	Signal Amplification Strategy	Detection Range	Limit of Detection	Ref.
Bisphenol A	$Ru(phen)_3{}^{2+}$ can integrate into the grooves of HCR products (dsDNA)	2.0–50 × 10^3 pM	1.5 pM	[28]
cTnI	Au nanoclusters and HCR signal amplification	5–5 × 10^4 fg/mL	1.01 fg/mL	[29]
Human immunodeficiency virus DNA	Target DNA triggered RCA signal amplification (RCA)	100–1 × 10^8 aM	27.0 aM	[30]
HPV DNA	Bovine serum albumin carrier platforms and hyperbranched rolling circle amplification	10–1.5 × 10^4 fM	7.6 fM	[31]
Hg^{2+}	Exonuclease III-assisted CRISPR/Cas12a	0–1 × 10^6 fM	0.45 fM	[32]
BT63DNA	ExoIII enzyme-assisted hybridization chain reaction combined with nanoparticle-loaded multiple probes	0.1–1 × 10^4 fM	0.036 fM	[33]
MicroRNA	A synergistic promotion strategy for 3D DNA walker amplification	10–1 × 10^8 aM	2.9 aM	[34]
miRNA-141	3D DNA walker-assisted CRISPR/Cas12a trans-cleavage	1–1 × 10^7 fM	0.33 fM	[35]
SARS-CoV-2	Target DNA-participated entropy-driven amplified reaction	1–1 × 10^5 fM	2.67 fM	[36]
MicroRNA let-7a	Swing arm location-controllable DNA walker based on the DNA tetrahedral nanostructures (DTNs)	10–1 × 10^8 fM	4.92 fM	[37]
8-hydroxy-2′-deoxyguanosine	Target-induced multi-DNA release and nicking enzyme amplification strategy	100–1 × 10^7 fM	25 fM	[38]
ochratoxin A	Nicking endonuclease-powered DNA walking machine	0.05–5 nM	0.012 nM	[39]
Myocardial miRNA	DNAzyme-regulated resonance energy transfer	10–1 × 10^7 fM	2.44 fM	[40]
carcinoembryonic antigen	DNAzyme-driven DNA walker amplification	1–1 × 10^8 fg/mL	0.21 fg/mL	[41]
5-Hydroxymethylcytosine	DNAzyme motor triggered by strand displacement amplification	1–1 × 10^6 fM	0.49 fM	[42]
MircoRNA-21	Localized DNA cascade reaction (LDCR) in a DNA nanomachine	100–1 × 10^9 aM	10.7 aM	[43]

2.1. Enzyme-Assisted DNA Amplification Strategies

As they are a type of enzyme, polymerases can catalyze DNA and RNA synthesis. They can replicate DNA and form long, linear, tandem, or repetitive chains of DNA with the assistance of a polymerase enzyme from the DNA template, primers, and deoxyribonucleoside triphosphate (dNTP) [44]. Polymerase chain reaction (PCR) remains the traditional and the "gold standard" enzyme-assisted DNA amplification strategy in bioanalysis due to its high sensitivity and low cost [45]. However, it has significant disadvantages, including the requirement of sophisticated and complicated processes and the presence of false-positive signals, which limit its practical use in the ECL domain. As alternative polymerase-based amplification techniques, rolling circle amplification (RCA) and hyperbranched RCA (HRCA) have attracted more attention, as they not only inherit isothermal amplification, but also promote improving the amplification efficiency. RCA requires a circular probe and DNA or RNA primers. In the presence of polymerases, prolonged extended ssDNA or double-stranded DNA is synthesized from the primer and the circular probe, and the ECL signal is enhanced based on the RCA product that it is loaded with or based on the in situ form abundant in the ECL luminophores. For example, as shown in Figure 1, two hairpin DNAs (H1 and H2) hybridize with the target DNA, releasing two output DNAs (W1 and W2) with the aid of exonuclease III. Subsequently, the concatenated DNA structure can be opened and can unlock the padlock oligonucleotide probe after being hybridized with two output DNAs. Under the action of the T4 DNA ligase, padlock DNA and other DNA primers are ligated to initiate the RCA reaction. Afterward, a prolong-extended ssDNA complementary to ruthenium (Ru)-labeled ssDNA is produced. Therefore, massive ruthenium (Ru)-labeled ssDNA is captured by the RCA products to generate a remarkable ECL signal, resulting in a large increase in amplification efficiency. The biosensor shows the highly specific and ultrasensitive detection of human immunodeficiency virus (HIV) DNA fragments when the detection limit is down to 27.0 aM [30]. Yen et al. utilized the RCA strategy to produce massive, long ssDNAs with a pH-dependent i-motif forming sequence, which was able to bind with hemin and catalyze the ECL reaction with the assistance of luminol/H_2O_2 solution. Because the i-motif structure is sensitive to pH change, a novel solid-state sensor for pH detection with a wide dynamic range from pH 4.0 to 7.4 was proposed [46]. To further improve the reaction efficiency, He et al. fabricated a high-reproducibility-and-sensitivity ECL biosensor for human papillomavirus 16 E6 and E7 by employing bovine serum albumin as a carrier platform to improve local steric hindrance and the HRCA strategy. After the addition of the target HPV DNA, HRCA occurred. Abundant HRCA products with double-stranded DNA (dsDNA) fragments of different lengths were generated, providing enough double helix space for the insertion of dichlorotris (1,10-phenanthroline) ruthenium(II) hydrate $[Ru(phen)_3]^{2+}$, which was acting as an ECL indicator, and releasing a strong and easily detected ECL signal [31]. Most RCA or HRCA-based ECL biosensors have shown great potential to avoid false-positive signals while also improving the sensitivity.

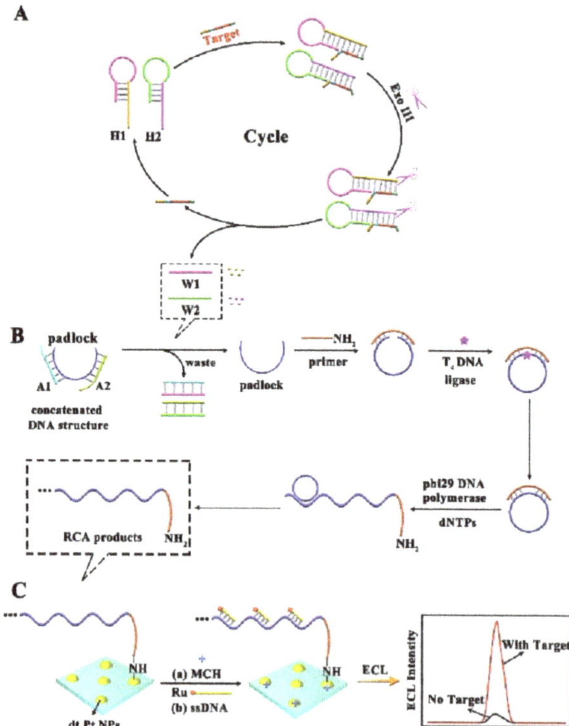

Figure 1. Example of the sensitive biosensor explored for HIV DNA fragment detection: (**A**) double recognition-triggered single-target recycling and (**B**) concatenated DNA structure-controlled rolling circle amplification; (**C**) construction process of the proposed ECL. Reproduced with permission from [30]. Copyright 2021, American Chemical Society.

Another commonly used enzymatic amplification technology in the ECL domain is cleaving enzyme-assisted amplification, which can be divided into two types according to the function of the enzyme used: (1) endonucleases, including the nicking endonuclease (NEase) and duplex-specific nuclease and (2) exonucleases, including exonuclease I, exonuclease III, and T7 Exo. Cleaving enzymes are a class of enzymes that preferentially cleave the phosphodiester bonds of nucleic acids. The released DNA is recycled in the next round, leading to multiple cycles of signal amplification induced by multiple capture and release cycles of the target [39,47]. Zhao et al. fabricated an aptasensor for 8-hydroxy-2′-deoxyguanosine detection for early diagnosis via a target-induced multi-DNA release and nicking enzyme amplification strategy. Aptamers on magnetic beads were hybridized with three kinds of short DNA. After the aptamer specifically recognized the target, the three kinds of short DNA were released, and three-fold signal amplification increases were obtained. Following that, the Fc-labeled DNA hybridized with the released DNA with the help of a nicking endonuclease (Nt.AlwI). The substrate strand (Fc-HP) was cleaved into two parts, and the Fc-labeled DNA could then leave the surface of the electrode, resulting in a strong ECL intensity. At the same time, the three kinds of short DNA were released again and were reused to initiate the repeated hybridization–cleavage cycles because of the nicking endonuclease-assisted recycling amplification [38].

DNAzyme is a functional DNA molecule that shows catalytic activity similar to that of traditional protein enzymes. DNAzyme normally contains a substrate strand containing an embedded cleavage site and a binding site for metal ions (e.g., Pb^{2+}, Cu^{2+}, Mn^{2+}, and Zn^{2+}). The binding of metal ions triggers the catalytic activity of DNAzyme and the subsequent splitting of the substrate strand [40,48–52]. Therefore, as a metal ion-dependent enzyme, DNAzyme has been incorporated into sensors for the detection of various metal ions. Currently, DNAzymes are usually designed as a trigger to release the initiator DNA sequence, with the potential to further initiate other amplification reactions [53]. Therefore, DNAzymes are often integrated with other DNA amplification strategies to develop multiple-signal amplification-based biosensors, which can further improve ECL performance. By combining DNAzyme with cascading amplification, Sun et al. [54] developed an ultrasensitive and multi-targeted ECL sensing platform for the analysis of myocardial miRNAs. Three myocardial miRNAs were successfully detected to have a detection limit as low as 29.6 aM.

Despite the participation of enzymes greatly improving the sensitivity of biosensors, enzymes require strict experimental conditions to maintain the catalytic activity, with examples of conditions including pH and temperature. Furthermore, the involvement of enzymes increases the experimental costs and the number of complex procedures.

2.2. Enzyme-Free Amplification Strategies

Various nonenzymic DNA amplification technologies, such as hybridization chain reaction, catalyzed hairpin assembly [55], and entropy-driven catalysis [36], have been used for the fabrication of enzyme-free biosensing platforms. Among these techniques, hybridization chain reaction (HCR) has been widely integrated into ECL biosensor development to construct nonenzymatic DNA biosensors [56]. Many reviews focused on traditional linear HCR assembly, and novel forms of HCR have been discussed [57]. Traditional HCR assembly requires single-strand initiator DNA and two harpin fuel DNAs. Single-strand initiator DNA hybridizes with the two fuel DNAs to form a long linear double-stranded DNA polymer under mild conditions, enabling more ECL reagents such as $[Ru(phen)_3]^{2+}$ to be embedded into the dsDNA or more ECL reagent-labeled DNA to be bound to the DNA, resulting in remarkable ECL improvement [58,59]. However, HCR's reaction efficiency is low due to its restricted presence on the electrode surface, which limits the efficiency of signal amplification. Lin's group proposed a sensitive electrochemiluminescence biosensor based on a click chemistry-triggered hybridization chain reaction for pyrophosphatase detection in a homogeneous solution. The hybridization chain reaction was processed in a homogeneous solution, which obviously improved the amplification efficiency [27]. With the development of molecular programming, diverse nucleic acid reaction circuits based on HCR have been proposed to form branched or dendritic nanostructures. Lin's group successfully integrated branched hybridization chain reaction (BHCR) into a biosensing approach (Figure 2). BHCR, which is a derivative of HCR, possesses multiple unique reaction orientations that greatly accelerate the reaction and improve the amplification efficiency. Therefore, BHCR not only inherits the properties of the nonenzymatic and isothermal amplification characteristics of HCR. However, it also possesses the advantages of rapid reaction kinetics and a high amplification efficiency because of its unique multiple reaction orientations [60].

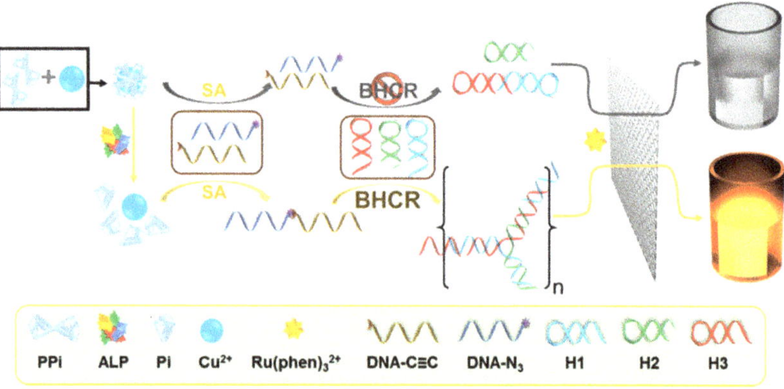

Figure 2. Example of the ECL biosensor for ALP detection based on BHCR amplification. Reproduced with permission from [60]. Copyright 2021, American Chemical Society.

Additionally, benefiting from DNA molecular programming, nanomachine-based amplification strategies have started to receive attention. Various DNA structures with different functional properties, such as DNA motors, DNA robots, and DNA walkers, have been designed and exploited for the construction of high-sensitivity biosensors. Especially, DNA nanomachines can be integrated for the quantitative detection of biomolecules using different amplification strategies [61]. As an example, DNA walkers, which are a type of autonomous nanomachine, show broad special cascade signal enhancement characteristics that are generated by their autonomous movement along the designed track. To date, most DNA walkers based on one walking strand are referred to as a single-leg DNA walkers; therefore, the walking area is limited, and the immediate response is slow. To overcome the above issue, some efforts have been devoted to the development of ECL strategies using multipedal DNA walkers by attaching multiple walking strands. It is worth mentioning that DNA walkers normally need to be driven by other strategies, such as DNAzyme, endonuclease-mediated hydrolysis, or toehold-mediated strand displacement (TMDR). As shown in Figure 3, Wang et al. adopted catalytic hairpin assembly (CHA) to trigger a tripedal DNA walker in the presence of miRNA-21. The tripedal DNA walker formed a Y-shaped structure with three "legs" and walked on a DNA track, producing a significant ECL signal. Combining CHA and tripedal DNA walker to develop a dual signal amplification strategy, high-sensitivity miRNA-21 detection was achieved, with a superior detection limit of 4 aM and a broad linear range of 10 aM to 1 pM [22].

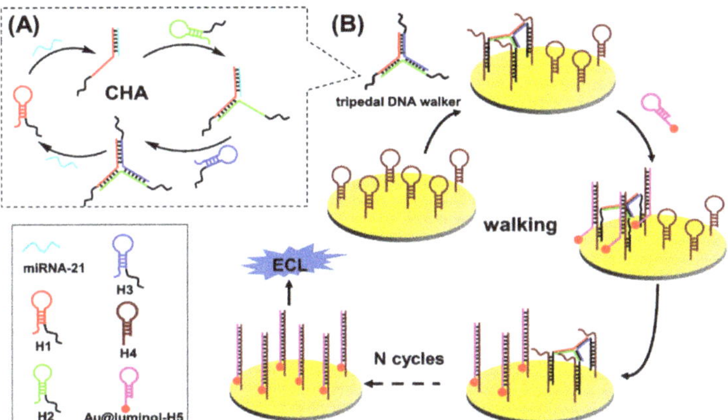

Figure 3. Example of the ECL biosensor for miRNA-21 detection based on the DNA walker strategy: (**A**) generation process of the tripedal DNA walker; (**B**) walking cycles of the DNA walker. Reproduced with permission from [22]. Copyright 2020, American Chemical Society.

3. Efficient ECL Luminophores

For now, most of the reported ECL biosensors and all commercial ECL assays are developed based on the ECL luminescent reagent tris(2,2′-bipyridineruthenium(II) ($[Ru(bpy)_3]^{2+}$) [22,62]; however, the quantum yield of $[Ru(bpy)_3]^{2+}$ is low, which can limit the sensitivity of the assays. Cyclometalated iridium complexes, which exhibit higher luminescence efficiency and long excited state lifetimes, have been reported as alternative luminescent reagents for $[Ru(bpy)_3]^{2+}$. The ECL efficiency from $(pq)_2Ir(acac)$ is reported to be significantly higher than that of $[Ru(bpy)_3]^{2+}$ [63]. Cyclometalated iridium complexes also allow the ECL emission to be tuned from the visible region to the UV region, which promotes the appearance of resolvable "mixed-ECL" from solutions containing multiple luminophores [64]. However, the poor water solubility of iridium complexes limits its practical use.

In the past few decades, experimental and theoretical studies have shown that quantum dots possess unique ECL properties; however, the ECL intensity of QD is unstable and weak [65]. To meet the demands for efficient luminophores, nanocarriers loaded with QDs demonstrate excellent ECL performance, with signals being obviously amplified. In general, silica nanoparticles, carbon nanomaterials, metal organic frameworks (MOFs), and transition metal oxides are applied as nanocarriers to load QDs. To date, QD-based biosensors have been widely used for ultrasensitive analysis [66–69], which benefits from their unique features such as their excellent biocompatibility, water solubility, and low toxicity. Zhuo's group [70] prepared IRMOF−3 accelerator-enriched QDs (CdTe@IRMOF−3@CdTe) using a direct encapsulation method for the trace detection of cTnI. With $S_2O_8{}^{2-}$ as the coreactant, the composites showed enhanced ECL intensity compared to other QD aggregates. Moreover, IRMOF−3 can be functioned as a coreactant accelerator that can further amplify the ECL signal. The developed sensor exhibited a wide dynamic range of 1.1 fg/mL to 11 ng/mL for cTnI, with a limit of detection (LOD) of 0.46 fg/mL (Figure 4).

Recently, certain metal nanoclusters have also been found to possess ECL properties [71]. To date, gold nanoclusters have emerged as a new class of ECL emitters due to their stable optical and electrochemical properties, monodispersity size, and low- and well-defined band gap and high atomic accuracy [72]. Nie et al. [73] reported a novel luminophore, a Au NC-based metal-organic framework (Au NC-based MOF) that showed 10-fold-enhanced anodic ECL efficiency over aggregated GSH-Au NCs in an aqueous solution, and when rutin was the model analyte, a low detection limit of 10 nM was achieved.

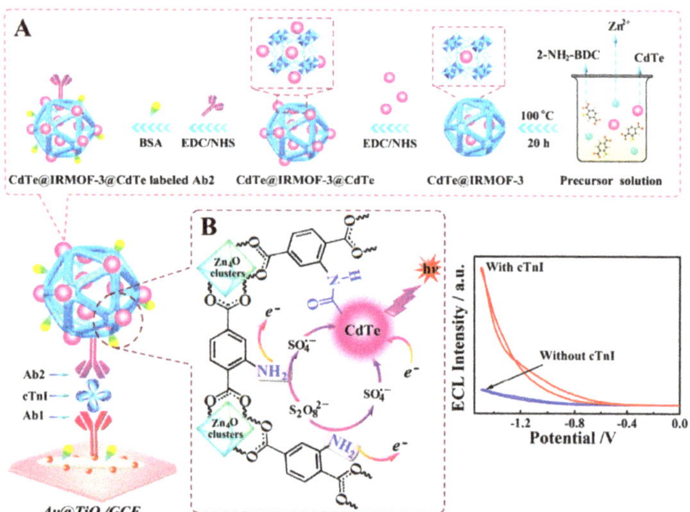

Figure 4. (**A**) Preparation of the signal probe. (**B**) Possible mechanism of IRMOF−3 accelerator-mediated enhancement of cTnI detection in the CdTe/$S_2O_8{}^{2-}$ system. Reproduced with permission from [70]. Copyright 2018, American Chemical Society.

Other types of nanomaterials have recently been used in ECL bioanalysis, such as graphite-phase carbon nitride (CN) [74], a novel nitrogen-rich two-dimensional carbon material, which has a wide range of applications in photocatalysis and biosensing [75,76]. Surface-modified CN is of great significance due to its practical application, especially for the regulation of its ECL characteristics [77,78]. Ji et al. [79] conducted a systematic scientific investigation on the application of graphite-phase carbon nitride (CN) in the ECL field. They destroyed the stacking between layers of CN using a simple mechanical grinding method to obtain CN of a smaller size. The prepared CN nanocrystals not only retained the photoelectric characteristics of the original CN but were also able to overcome the CN defects on the surface of the materials, which provided a basis for its further application in the field of biosensors.

In total, various kinds of nanomaterials, including carbon-based nanomaterials, metal nanomaterials, and other inorganic or organic nanomaterials, as have been used as coreactant or luminophores to enhance ECL performance in biosensing. However, the reported mechanism proposed that the ECL of nanomaterials be generated through both the electron and holes when the dots are injected separately onto the surface. The surface states of nanomaterials are reflected in the ECL performance. The ECL response can be increased by increasing the efficiency of electron–hole recombination via different strategies.

The ECL is much more sensitive to the surface states of CDs. Thus, developing unique luminophores and especially nanomaterials that enhance ECL efficiency as well as enable sensitive targets sensing remains a challenging research area, and the development of advanced nanomaterials is an important research direction in the design of amplified ECL biosensors.

4. Ratiometric Strategies

4.1. Potential-Resolved Ratiometric Strategies

Potential-resolved strategies require two ECL emitters that emit light at different potentials [80]. The difference in the peak emission potential between the two ECL peaks should be enough to be easily distinguished without interference [81]. Based on the intensity ratio of the two ECL peaks at different potentials, many ratiometric biosensors have been established for the detection of prostate-specific antigen (PSA), aflatoxin B1

(AFB1), dopamine, miRNA, metal cations, and cancer cells, which not only improve the sensitivity, but also avoid false-positive errors in complex matrices [24,82]. However, this kind of ratiometric biosensor also requires two coreactants in addition to additional luminophores [44,83,84], which increase the complexity of the whole experimental process. Some efforts have been made to develop single-luminophore-based ratiometric sensors [24]. As shown in Figure 5, Cui's group fabricated a ratiometric biosensor for the determination of miR-133a. The novel ECL luminophores, graphitic carbon nitrides ($g\text{-}C_3N_4$) functionalized by N-(aminobutyl)-N-(ethylisoluminol) (ABEI) ($g\text{-}C_3N_4$/ABEI), were used as the sensing interface. $g\text{-}C_3N_4$ emitted ECL light in the negative potential range at −1.6, while the ECL emission of ABEI occurred in the positive potential range at +1.2 in the presence of H_2O_2. Based on the ECL intensity ratio of the $g\text{-}C_3N_4$/ABEI, the proposed bioassay presented a linear range from 0.1 fM to 1.0 pM and a LOD of 48.0 aM. Cao's group fabricated a potential-resolved ECL biosensor based on novel nano-luminophores: nano-graphene oxide wrapped titanium dioxide (nGO@TiO_2NLPs), which showed potential-resolved ECL properties in a neutral aqueous solution using $K_2S_2O_8$ as a coreactant. These new luminophores could also be employed as luminescent cathode and anode materials simultaneously. In the presence of cardiac troponin I (cTnI), the aptamer protrudes from the electrode surface owing to its rigidity, leading to a reduction in the charge transfer resistance of the modified working electrode and a ratio enhancement of the two ECL signals of the nGO@TiO_2 NLPs. According to the increased ECL ratio, cTnI was quantified using the ratiometric ECL aptasensor, with a linear dynamic range of 1.0×10^{-13}–1.0×10^{-10} mol/L.

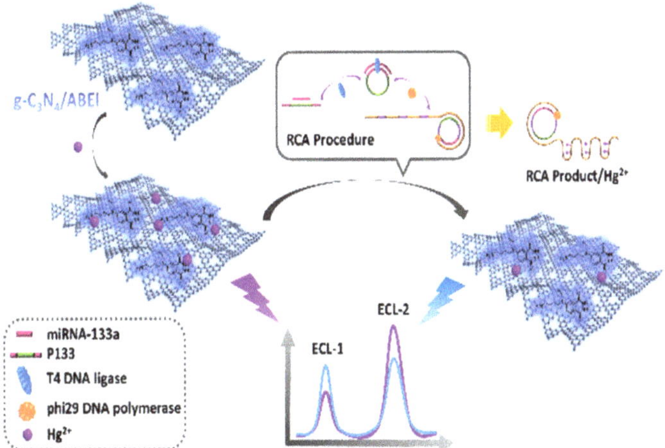

Figure 5. Potential-resolved ratiometric strategy for the detection of miRNA-155. Reproduced with permission from [25]. Copyright 2020, American Chemical Society.

4.2. Spectrum-Resolved Ratiometric Strategies

Compared to potential-resolved sensors, spectrum-resolved ECL sensors not only inherit the advantages of high accuracy but can also be used in a narrow potential range [85,86]. These sensors require ECL emitters that emit light at different wavelengths [87], and there should also ideally be overlap between the ECL emission and absorption peaks to allow resonance energy transfer (RET) to occur. For example, based on ECL RET, a dual-wavelength ratiometric ECL sensor was established for the detection of the amyloid-β protein in human serum [88]. The process is displayed in Figure 6. A couple of emitters, Ru@TiO_2@Au nanomaterial/gold nanoparticle (AuNP)-modified graphitic carbon nitride nanosheets ($g\text{-}C_3N_4$NSs), were employed as the energy receptor and energy donor. With the addition of a target protein, due to the RET effect between the two emitters, the $g\text{-}C_3N_4$NSs signal at about 460 nm was gradually quenched, while the signal of Ru@TiO_2@Au increased at about 620 nm. During the experiment, based on the ratio of I_{460nm}/I_{620nm}, the linearity range of Aβ42 was found to be from 1×10^{-5} to 200 ng/mL, with a limit of detection (LOD) of 2.6 fg/mL, which makes the sensor suitable for application in clinical diagnosis.

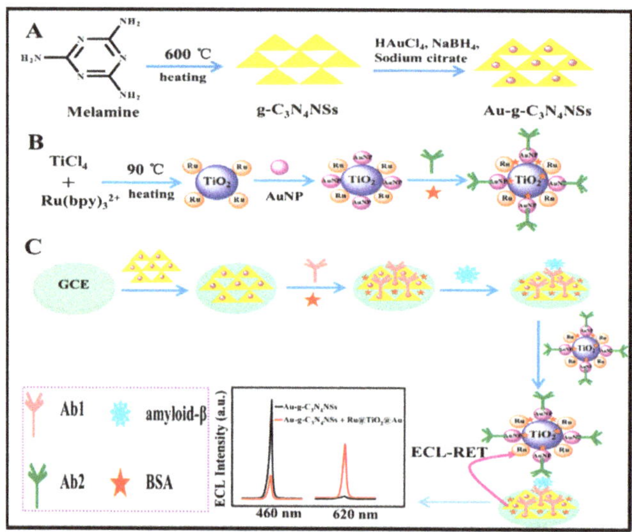

Figure 6. (**A**) Preparation of Au-g-C_3N_4NSs; (**B**) fabrication of Ru@TiO_2@Au-Ab2 conjugate; and (**C**) assembly process of DWR-ECL immunosensor. Reproduced with permission from [88]. Copyright 2021, American Chemical Society.

5. Surface-Enhanced Strategies

Previous reports have demonstrated that the localized surface plasmon resonance (LSPR) of metal nanoparticles (such as gold and silver) can significantly enhance the spectral signal [89,90]. LSPR is a physical phenomenon that is generated when the surface plasma of noble metal nanoparticles is irradiated by incident light with the same frequency [91]. LSPR can generate a local electromagnetic field around noble metal nanoparticles, thus enhancing the spectral signal [92]. By controlling the distance between the surface of noble metal nanoparticles and the ECL luminophore, the intensity of ECL can be greatly improved [93]. This phenomenon is called surface-enhanced electrochemiluminescence (SEECL). As shown in Figure 7, credible evidence of this process as well as a detailed mechanism of it has been presented by Wang and co-workers [94]. The ECL signal of the Au NP@SiO_2-modified electrode was 10 times higher than that of the bare electrode. Since this initial report on SEECL, a series of ultra-sensitive biosensors based on SEECL have been developed [90,93,95–98]. For example, an ultra-sensitive biosensor for Hg^{2+} using the local surface plasmon resonance (LSPR) of gold nanorods (Au NR) has been proposed. When Hg^{2+} is present, the conformation of ssDNA probes changed into a hairpin structure by forming a T-Hg^{2+}-T structure. $[Ru(bpy)_3]^{2+}$ can be embedded into a hairpin DNA probe to generate ECL emissions, while the LSPR of Au NRs can enhance ECL emissions. As the Hg^{2+} concentration increases, the ECL intensity also increases, and the detection limit of the sensor reaches 10 fM [99]. The LSPR of metal nanoparticles is also often used to enhance the electrochemiluminescence of quantum dots. CuZnInS quantum dots are a novel ECL luminescent material, but they suffer from a low ECL efficiency. Based on CuZnInS quantum dots (QDs) and gold nanoparticles (AuNPs), Chen et al. [97] developed a novel DNA electrochemiluminescence sensor for the highly sensitive detection of the epidermal growth factor receptor (EGFR) gene closely related to lung cancer. The detection range of EGFR ranged from 0.05 to 1 nmol/L, and the detection limit was 0.0043 nmol/L. Apart from Au monomers, Au–Au dimers can be applied in the construction of biosensors to further improve the signal intensity and the sensitivity due to the surface plasmon coupling between two Au monomers, resulting in high electromagnetic field enhancement [100]. It is worth noting that the metallic substrates used in most surface-enhanced strategies are limited to Au nanomaterials. Although Au nanomaterials are easily synthesized and have good stability, other nanomaterial candidates should be explored for the purposes of determining additional excellent plasmonic properties. Therefore, it is worth noting that Ag nanomaterials have also shown excellent plasmonic properties. Cao et al. employed AgNP nanocrystals as the LSPR source and MoS_2 QDs as the ECL emitter, which demonstrated that Ag nanomaterials can greatly enhance the effects on ECL. The detection limit for microRNA-21 was 0.2 fM.

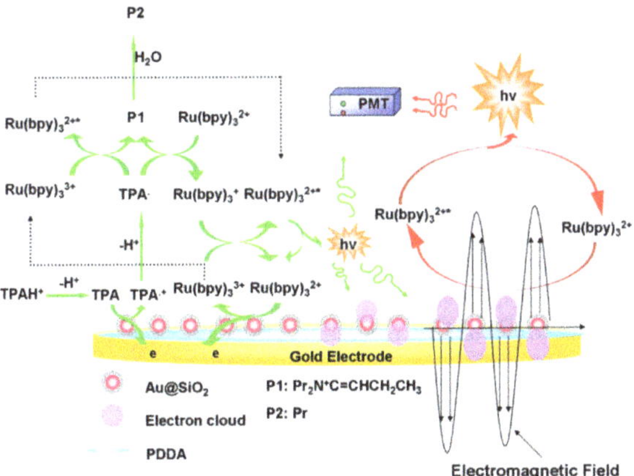

Figure 7. The proposed SEECL mechanism. Reproduced with permission from [94]. Copyright 2015, Scientific Reports.

In addition, imaging technology coupled with surface-enhanced electrochemiluminescence has been used in biosensing [14,101]. Liu and his team [102] constructed a biosensor for the determination of *Escherichia coli* through the mode conversion between resonance energy transfer (ECL-RET) and surface plasmon-coupled ECL (SPC-ECL) by controlling the distance between the BN QDs and Au NPs. The BN QDs and Au NPs were separately modified at the two ends of the hairpin DNA. When there was no target, resonance energy transfer occurred due to the close distance between the BN QDs and Au NPs, and the ECL signal of the BN QDs was inhibited by the AuNPs. In the presence of the target, the hairpin DNA hybridized with the target, and the structure changed to a linear conformation; therefore, as the distance between the BN QDs and Au NPs increased, the ECL intensity also increased. As a result, ECL-RET is replaced by the SPC-ECL effect, and the ECL signal is enhanced. In addition, the author presents the ECL signal for the first time using a CMOS camera imaging method and a smartphone. The fabricated biosensor showed a LOD of 0.3 pmol/L, with a linear range from 1 pmol/L to 5 nmol/L.

Compared to traditional biosensors, biosensors based on surface-enhanced strategies are simpler and more sensitive and show great potential for the detection of ultra-trace biomarkers for clinic diagnostics. Moreover, biosensors can be more biocompatible owing to the choice of Au or Ag nanomaterials as an LSPR source. However, real sample analysis is still a significant challenge.

6. Other Types of Signal-Amplification Strategies

Most biosensors mainly depend on single-signal output, which make them easily susceptible to other intrinsic and extrinsic factors. Single-signal assays in particular cannot meet the increasing demands for clinical diagnosis because some diseases are related to one or more biomarkers. Therefore, many efforts have been devoted to developing ECL biosensors based on multiple-signal outputs, which enable multiple analytes to be measured simultaneously. For example, Su's group [64] successfully synthesized three novel ECL emitters, including ruthenium and iridium complexes. Based on the ECL emitters with different spectral peaks at different potentials, a multiplex immunoassay for the simultaneous detection of carcinoembryonic antigen (CEA), alpha-fetoprotein (AFP), and beta-human chorionic gonadotropin (β-HCG) was developed. However, it is challenging to develop novel luminophores with potential-resolved or spectrum-resolved properties because the number of luminophores pairs for multiplex immunoassays is limited.

It is well-known that the coreactant process is one of the main mechanisms of ECL. Thus, the coreactant plays an important role during signal transduction, and the development of novel and efficient coreactants is another amplification strategy for biosensor construction. Nanomaterials, which are emerging as a new class of coreactants, have been exploited as attractive candidates for inorganic molecules, especially carbon dot-based nanomaterials. For instance, Wang' group [103] proved that the use of boron nitride quantum dots as a coreactant to enhance the ECL of $[Ru(bpy)_3]^{2+}$ resulted in a 400-fold enhancement being achieved. This is due to amino-bearing groups on the surface of the quantum dots. Thus, enhanced ECL efficiency is highly related to the surface state of nanomaterials [104].

7. Conclusions and Perspectives

Signal-amplification-based biosensors have gained great attention and have undergone rapid development owing to the requirements for ultrasensitive biosensors and trends towards early clinical diagnosis. Signal amplification strategies open novel approaches for developing ultrasensitive bioassays with a wide dynamic range. Specifically, they offer opportunities for monitoring early diagnosis, monitoring disease progression, and predicting disease in biomedical diagnoses. Among these strategies, DNA-assisted techniques are the most popular for signal amplification during ECL bioassays due to the advantages of specific base pairing, programmable operation, and predictable assembly. Enzyme-assisted DNA amplification strategies have achieved enhanced sensitivity in ECL, but enzymatic reactions are susceptible to environmental factors, which ultimately affect the DNA amplification efficiency and limit their application in complex biological systems. Therefore, the development of low-cost, sensitive, and enzyme-free strategies is the research direction to achieve future commercialization. However, the commercialization of point of care testing is still in its early stages. Multiple DNA circuits make the processes more complicated, and the amplification efficiency at each step is still unknown, affecting the detection accuracy. Therefore, highly efficient luminophores have been explored to amplify ECL signals. Finally, ratiometric strategies and surface-enhanced ECL approaches have been introduced into biosensing, which can not only improve the sensitivity, but can also increase the accuracy. This review presents the recent progress in ECL biosensors that have been integrated with various kinds of signal amplification strategies, hoping to provide guidance for designing novel ECL biosensors.

Based on the above-mentioned strategies, ECL biosensors have the ability to achieve the ultra-trace level detection of targets. However, most biosensors are only demonstrated to be successful in their principle of concept and the translation of research into industrial manufacturing and marketing remains a significant challenge. To achieve commercialization, there is urgency to develop disposable, low-cost, and lab-on-chip ECL platforms to conduct tests with a rapid response time and that have high convenience. Specifically, ECL systems that are autonomous and miniaturized and that show great potential in practical applications represent future research trends. In addition, the development of various low-cost photodetectors such as CCD cameras, smartphone cameras, and other imaging techniques coupled with ECL biosensors can realize multi-component analysis and single-molecule detection. Single-biomolecule ECL imaging should be the subject of major efforts to greatly expand ECL applications in bioanalysis.

Author Contributions: Conceptualization, L.C., Y.Y. and Y.W.; writing, Y.H. and L.C.; supervision, L.G., L.L. and Y.Z. All authors have read and agreed to the published version of the manuscript.

Funding: This work was financially supported by the National Natural Science Foundation of China (22074054), the key Research and Development Program of Zhejiang Province (2020C02022), the Zhejiang Provincial High-level Talent Special Support Plan (2021R52044), and the Natural Science Foundation of Zhejiang Province (LQ20B050002, LQ20B050004, LGF22B050005).

Institutional Review Board Statement: Not applicable.

Informed Consent Statement: Not applicable.

Data Availability Statement: Not applicable.

Conflicts of Interest: The authors declare no conflict of interest.

References

1. Alarfaj, N.A.; El-Tohamy, M.F.; Oraby, H.F. New immunosensing-fluorescence detection of tumor marker cytokeratin-19 fragment (CYFRA 21-1) via carbon quantum dots/zinc oxide nanocomposite. *Nanoscale Res. Lett.* **2020**, *15*, 1–14. [CrossRef] [PubMed]
2. Su, M.; Liu, S. Solid-state electrochemiluminescence analysis with coreactant of the immobilized tris (2, 2′-bipyridyl) ruthenium. *Anal. Biochem.* **2010**, *1*, 1–12. [CrossRef] [PubMed]
3. Zhu, L.; Ye, J.; Yan, M.; Zhu, Q.; Yang, X. A wavelength-resolved electrochemiluminescence resonance energy transfer ratiometric immunosensor for detection of cardiac troponin I. *Analyst* **2019**, *144*, 6554–6560. [CrossRef]
4. Neves, M.M.; González-García, M.B.; Hernández-Santos, D.; Fanjul-Bolado, P. A miniaturized flow injection analysis system for electrogenerated chemiluminescence-based assays. *ChemElectroChem* **2017**, *4*, 1686–1689. [CrossRef]
5. Forry, S.P.; Wightman, R.M. Electrogenerated chemiluminescence detection in reversed-phase liquid chromatography. *Anal. Chem.* **2002**, *74*, 528–532. [CrossRef]
6. Sun, S.; Wei, Y.; Wang, H.; Cao, Y.; Deng, B. A novel electrochemiluminescence sensor coupled with capillary electrophoresis for simultaneous determination of quinapril hydrochloride and its metabolite quinaprilat hydrochloride in human plasma. *Talanta* **2018**, *179*, 213–220. [CrossRef]
7. Huang, Y.; Lu, Y.; Huang, X.; Wang, J.; Qiu, B.; Luo, F.; Lin, Z. Design of an electrochemiluminescence detection system through the regulation of charge density in a microchannel. *Chem. Sci.* **2021**, *12*, 13151–13157. [CrossRef]
8. Hercules, D.M. Chemiluminescence resulting from electrochemically generated species. *Science* **1964**, *145*, 808–809. [CrossRef]
9. Santhanam, K.; Bard, A.J. Chemiluminescence of electrogenerated 9, 10-Diphenylanthracene anion radical1. *J. Am. Chem. Soc.* **1965**, *87*, 139–140. [CrossRef]
10. Lee, W.-Y. Tris (2, 2′-bipyridyl) ruthenium (II) electrogenerated chemiluminescence in analytical science. *Microchim. Acta* **1997**, *127*, 19–39. [CrossRef]
11. Fähnrich, K.A.; Pravda, M.; Guilbault, G.G. Recent applications of electrogenerated chemiluminescence in chemical analysis. *Talanta* **2001**, *54*, 531–559. [CrossRef]
12. Liu, Z.; Qi, W.; Xu, G. Recent advances in electrochemiluminescence. *Chem. Soc. Rev.* **2015**, *44*, 3117–3142. [CrossRef] [PubMed]
13. Hai-Juan, L.; Shuang, H.; Lian-Zhe, H.; Guo-Bao, X. Progress in Ru $(bpy)_3{}^{2+}$ electrogenerated chemiluminescence. *Chin. J. Anal. Chem.* **2009**, *37*, 1557–1565.
14. Wang, Q.; Ren, Z.-H.; Zhao, W.-M.; Wang, L.; Yan, X.; Zhu, A.-s.; Qiu, F.-m.; Zhang, K.-K. Research advances on surface plasmon resonance biosensors. *Nanoscale* **2022**, *14*, 564–591. [CrossRef] [PubMed]
15. Zhao, L.D.; Yang, X.; Zhong, X.; Zhuo, Y. Advances in Electrochemiluminescence Biosensors Based on DNA Walkers. *ChemPlusChem* **2022**, *87*, e202200070. [CrossRef]
16. Luo, W.; Chu, H.; Wu, X.; Ma, P.; Wu, Q.; Song, D. Disposable biosensor based on novel ternary Ru-PEI@ PCN-333 (Al) self-enhanced electrochemiluminescence system for on-site determination of caspase-3 activity. *Talanta* **2022**, *239*, 123083. [CrossRef]
17. Kurup, C.P.; Lim, S.A.; Ahmed, M.U. Nanomaterials as signal amplification elements in aptamer-based electrochemiluminescent biosensors. *Bioelectrochemistry* **2022**, *147*, 108170. [CrossRef]
18. Sun, T.; Du, J.; Li, Z.; Zhao, F. Recent Advancement in the development of hybridization chain reaction-based electrochemiluminescence biosensors. *Int. J. Electrochem. Sci.* **2022**, *17*, 2. [CrossRef]
19. Miao, P.; Chai, H.; Tang, Y. DNA Hairpins and Dumbbell-Wheel Transitions Amplified Walking Nanomachine for Ultrasensitive Nucleic Acid Detection. *ACS Nano* **2022**, *16*, 4726–4733. [CrossRef]
20. Yan, Y.; Li, J.; Li, W.; Wang, Y.; Song, W.; Bi, S. DNA flower-encapsulated horseradish peroxidase with enhanced biocatalytic activity synthesized by an isothermal one-pot method based on rolling circle amplification. *Nanoscale* **2018**, *10*, 22456–22465. [CrossRef]
21. Feng, Q.-M.; Guo, Y.-H.; Xu, J.-J.; Chen, H.-Y. Self-assembled DNA tetrahedral scaffolds for the construction of electrochemiluminescence biosensor with programmable DNA cyclic amplification. *ACS Appl. Mater. Interfaces* **2017**, *9*, 17637–17644. [CrossRef] [PubMed]

22. Wang, L.; Liu, P.; Liu, Z.; Zhao, K.; Ye, S.; Liang, G.; Zhu, J.-J. Simple tripedal DNA walker prepared by target-triggered catalytic hairpin assembly for ultrasensitive electrochemiluminescence detection of microRNA. *ACS Sens.* **2020**, *5*, 3584–3590. [CrossRef] [PubMed]
23. Oishi, M.; Saito, K. Simple single-legged DNA walkers at diffusion-limited nanointerfaces of gold nanoparticles driven by a DNA circuit mechanism. *ACS Nano* **2020**, *14*, 3477–3489. [CrossRef] [PubMed]
24. Shang, L.; Wang, X.; Zhang, W.; Jia, L.-P.; Ma, R.-N.; Jia, W.-L.; Wang, H.-S. A dual-potential electrochemiluminescence sensor for ratiometric detection of carcinoembryonic antigen based on single luminophor. *Sens. Actuators B Chem.* **2020**, *325*, 128776. [CrossRef]
25. Wang, J.; Haghighatbin, M.A.; Shen, W.; Mi, L.; Cui, H. Metal ion-mediated potential-resolved ratiometric electrochemiluminescence bioassay for efficient determination of miR-133a in early diagnosis of acute myocardial infarction. *Anal. Chem.* **2020**, *92*, 7062–7070. [CrossRef] [PubMed]
26. Li, Y.; Huang, C.Z.; Li, Y.F. Ultrasensitive electrochemiluminescence detection of MicroRNA via one-step introduction of a target-triggered branched hybridization chain reaction circuit. *Anal. Chem.* **2019**, *91*, 9308–9314. [CrossRef] [PubMed]
27. Huang, X.; Jia, J.; Lin, Y.; Qiu, B.; Lin, Z.; Chen, H. A highly sensitive electrochemiluminescence biosensor for pyrophosphatase detection based on click chemistry-triggered hybridization chain reaction in homogeneous solution. *ACS Appl. Mater. Interfaces* **2020**, *12*, 34716–34722. [CrossRef] [PubMed]
28. Zhang, H.; Luo, F.; Wang, P.; Guo, L.; Qiu, B.; Lin, Z. Signal-on electrochemiluminescence aptasensor for bi s p h e n o l A based on hybridization chain reaction and electrically heated electrode. *Biosens. Bioelectron.* **2019**, *129*, 36–41. [CrossRef]
29. Zhu, L.; Ye, J.; Yan, M.; Zhu, Q.; Wang, S.; Huang, J.; Yang, X. Electrochemiluminescence immunosensor based on Au nanocluster and hybridization chain reaction signal amplification for ultrasensitive detection of cardiac troponin I. *ACS Sens.* **2019**, *4*, 2778–2785. [CrossRef]
30. Zhang, X.; Zhou, Y.; Chai, Y.; Yuan, R. Double hairpin DNAs recognition induced a novel cascade amplification for highly specific and ultrasensitive electrochemiluminescence detection of DNA. *Anal. Chem.* **2021**, *93*, 7987–7992. [CrossRef]
31. He, Y.; Liu, Y.; Cheng, L.; Yang, Y.; Qiu, B.; Guo, L.; Wang, Y.; Lin, Z.; Hong, G. Highly reproducible and sensitive electrochemiluminescence biosensors for HPV detection based on bovine serum albumin carrier platforms and hyperbranched rolling circle amplification. *ACS Appl. Mater. Interfaces* **2020**, *13*, 298–305. [CrossRef] [PubMed]
32. Hang, X.-M.; Zhao, K.-R.; Wang, H.-Y.; Liu, P.-F.; Wang, L. Exonuclease III-assisted CRISPR/Cas12a electrochemiluminescence biosensor for sub-femtomolar mercury ions determination. *Sens. Actuators B Chem.* **2022**, *368*, 132208. [CrossRef]
33. Hai, H.; Chen, C.; Chen, D.; Li, P.; Shan, Y.; Li, J. A sensitive electrochemiluminescence DNA biosensor based on the signal amplification of ExoIII enzyme-assisted hybridization chain reaction combined with nanoparticle-loaded multiple probes. *Microchim. Acta* **2021**, *188*, 1–8. [CrossRef]
34. Yang, F.; Yang, F.; Tu, T.-T.; Liao, N.; Chai, Y.-Q.; Yuan, R.; Zhuo, Y. A synergistic promotion strategy remarkably accelerated electrochemiluminescence of SnO2 QDs for MicroRNA detection using 3D DNA walker amplification. *Biosens. Bioelectron.* **2021**, *173*, 112820. [CrossRef] [PubMed]
35. Wang, Q.; Liu, Y.; Yan, J.; Liu, Y.; Gao, C.; Ge, S.; Yu, J. 3D DNA walker-assisted CRISPR/Cas12a trans-cleavage for ultrasensitive electrochemiluminescence detection of miRNA-141. *Anal. Chem.* **2021**, *93*, 13373–13381. [CrossRef] [PubMed]
36. Fan, Z.; Yao, B.; Ding, Y.; Zhao, J.; Xie, M.; Zhang, K. Entropy-driven amplified electrochemiluminescence biosensor for RdRp gene of SARS-CoV-2 detection with self-assembled DNA tetrahedron scaffolds. *Biosens. Bioelectron.* **2021**, *178*, 113015. [CrossRef]
37. Liao, N.; Pan, M.-C.; Wang, L.; Yang, F.; Yuan, R.; Zhuo, Y. Swing arm location-controllable DNA walker for electrochemiluminescence biosensing. *Anal. Chem.* **2021**, *93*, 4051–4058. [CrossRef]
38. Zhao, R.-N.; Jia, L.-P.; Feng, Z.; Ma, R.-N.; Zhang, W.; Shang, L.; Xue, Q.-W.; Wang, H.-S. Ultrasensitive electrochemiluminescence aptasensor for 8-hydroxy-2′-deoxyguanosine detection based on target-induced multi-DNA release and nicking enzyme amplification strategy. *Biosens. Bioelectron.* **2019**, *144*, 111669. [CrossRef]
39. Wei, M.; Wang, C.; Xu, E.; Chen, J.; Xu, X.; Wei, W.; Liu, S. A simple and sensitive electrochemiluminescence aptasensor for determination of ochratoxin A based on a nicking endonuclease-powered DNA walking machine. *Food Chem.* **2019**, *282*, 141–146. [CrossRef]
40. Sun, Y.; Fang, L.; Han, Y.; Feng, A.; Liu, S.; Zhang, K.; Xu, J.-J. Reversible Ratiometric Electrochemiluminescence Biosensor Based on DNAzyme Regulated Resonance Energy Transfer for Myocardial miRNA Detection. *Anal. Chem.* **2022**, *94*, 7035–7040. [CrossRef]
41. Zhao, Y.; Tan, L.; Jie, G. Ultrasensitive electrochemiluminescence biosensor for the detection of carcinoembryonic antigen based on multiple amplification and a DNA walker. *Sens. Actuators B Chem.* **2021**, *333*, 129586. [CrossRef]
42. Li, X.-R.; Wang, L.; Liang, W.-B.; Yuan, R.; Zhuo, Y. Epigenetic Quantification of 5-Hydroxymethylcytosine Signatures via Regulatable DNAzyme Motor Triggered by Strand Displacement Amplification. *Anal. Chem.* **2022**, *94*, 3313–3319. [CrossRef] [PubMed]
43. Jiang, X.; Wang, H.; Chai, Y.; Li, H.; Shi, W.; Yuan, R. DNA cascade reaction with high-efficiency target conversion for ultrasensitive electrochemiluminescence microRNA detection. *Anal. Chem.* **2019**, *91*, 10258–10265. [CrossRef] [PubMed]
44. Wang, L.; Liu, P.; Liu, Z.; Cao, H.; Ye, S.; Zhao, K.; Liang, G.; Zhu, J.-J. A dual-potential ratiometric electrochemiluminescence biosensor based on Au@CDs nanoflowers, Au@luminol nanoparticles and an enzyme-free DNA nanomachine for ultrasensitive p53 DNA detection. *Sens. Actuators B Chem.* **2021**, *327*, 128890. [CrossRef]

45. Mason, J.T.; Xu, L.; Sheng, Z.-m.; O'Leary, T.J. A liposome-PCR assay for the ultrasensitive detection of biological toxins. *Nat. Biotechnol.* **2006**, *24*, 555–557. [CrossRef]
46. Ye, J.; Yan, M.; Zhu, L.; Huang, J.; Yang, X. Novel electrochemiluminescence solid-state pH sensor based on an i-motif forming sequence and rolling circle amplification. *Chem. Commun.* **2020**, *56*, 8786–8789. [CrossRef]
47. Li, D.; Li, Y.; Luo, F.; Qiu, B.; Lin, Z. Ultrasensitive homogeneous electrochemiluminescence biosensor for a transcription factor based on target-modulated proximity hybridization and exonuclease III-powered recycling amplification. *Anal. Chem.* **2020**, *92*, 12686–12692. [CrossRef]
48. Zhang, Y.; Xu, J.; Zhou, S.; Zhu, L.; Lv, X.; Zhang, J.; Zhang, L.; Zhu, P.; Yu, J. DNAzyme-triggered visual and ratiometric electrochemiluminescence dual-readout assay for Pb (II) based on an assembled paper device. *Anal. Chem.* **2020**, *92*, 3874–3881. [CrossRef]
49. Xia, M.; Zhou, F.; Feng, X.; Sun, J.; Wang, L.; Li, N.; Wang, X.; Wang, G. A DNAzyme-based dual-stimuli responsive electrochemiluminescence resonance energy transfer platform for ultrasensitive anatoxin-A detection. *Anal. Chem.* **2021**, *93*, 11284–11290. [CrossRef]
50. Wang, L.; Zhao, K.-R.; Liu, Z.-J.; Zhang, Y.-B.; Liu, P.-F.; Ye, S.-Y.; Zhang, Y.-W.; Liang, G.-X. An "on-off" signal-switchable electrochemiluminescence biosensor for ultrasensitive detection of dual microRNAs based on DNAzyme-powered DNA walker. *Sens. Actuators B Chem.* **2021**, *348*, 130660. [CrossRef]
51. Ni, J.; Zhang, H.; Chen, Y.; Luo, F.; Wang, J.; Guo, L.; Qiu, B.; Lin, Z. DNAzyme-based Y-shaped label-free electrochemiluminescent biosensor for lead using electrically heated indium-tin-oxide electrode for in situ temperature control. *Sens. Actuators B Chem.* **2019**, *289*, 78–84. [CrossRef]
52. Jiang, J.; Du, X. DNA-targeted formation and catalytic reactions of DNAzymes for label-free ratiometric electrochemiluminescence biosensing. *Talanta* **2021**, *225*, 121964. [CrossRef]
53. Gong, L.; Zhao, Z.; Lv, Y.-F.; Huan, S.-Y.; Fu, T.; Zhang, X.-B.; Shen, G.-L.; Yu, R.-Q. DNAzyme-based biosensors and nanodevices. *Chem. Commun.* **2015**, *51*, 979–995. [CrossRef] [PubMed]
54. Sun, Y.; Fang, L.; Zhang, Z.; Yi, Y.; Liu, S.; Chen, Q.; Zhang, J.; Zhang, C.; He, L.; Zhang, K. A multitargeted electrochemiluminescent biosensor coupling DNAzyme with cascading amplification for analyzing myocardial miRNAs. *Anal. Chem.* **2021**, *93*, 7516–7522. [CrossRef] [PubMed]
55. Zhang, Y.; Li, X.; Xu, Z.; Chai, Y.; Wang, H.; Yuan, R. An ultrasensitive electrochemiluminescence biosensor for multiple detection of microRNAs based on a novel dual circuit catalyzed hairpin assembly. *Chem. Commun.* **2018**, *54*, 10148–10151. [CrossRef]
56. Zhang, C.; Chen, J.; Sun, R.; Huang, Z.; Luo, Z.; Zhou, C.; Wu, M.; Duan, Y.; Li, Y. The recent development of hybridization chain reaction strategies in biosensors. *ACS Sens.* **2020**, *5*, 2977–3000. [CrossRef]
57. Chai, H.; Cheng, W.; Jin, D.; Miao, P. Recent progress in DNA hybridization chain reaction strategies for amplified biosensing. *ACS Appl. Mater. Interfaces* **2021**, *13*, 38931–38946. [CrossRef]
58. Augspurger, E.E.; Rana, M.; Yigit, M.V. Chemical and biological sensing using hybridization chain reaction. *ACS Sens.* **2018**, *3*, 878–902. [CrossRef]
59. Yang, D.; Tang, Y.; Miao, P. Hybridization chain reaction directed DNA superstructures assembly for biosensing applications. *TrAC Trends Anal. Chem.* **2017**, *94*, 1–13. [CrossRef]
60. Huang, X.; Bian, X.; Chen, L.; Guo, L.; Qiu, B.; Lin, Z. Highly sensitive homogeneous electrochemiluminescence biosensor for alkaline phosphatase detection based on click chemistry-triggered branched hybridization chain reaction. *Anal. Chem.* **2021**, *93*, 10351–10357. [CrossRef]
61. Zhang, C.; Ma, X.; Zheng, X.; Ke, Y.; Chen, K.; Liu, D.; Lu, Z.; Yang, J.; Yan, H. Programmable allosteric DNA regulations for molecular networks and nanomachines. *Sci. Adv.* **2022**, *8*, eabl4589. [CrossRef] [PubMed]
62. Sobhanie, E.; Salehnia, F.; Xu, G.; Hamidi, Y.; Arshian, S.; Firoozbakhtian, A.; Hosseini, M.; Ganjali, M.R.; Hanif, S. Recent trends and advancements in electrochemiluminescence biosensors for human virus detection. *TrAC Trends Anal. Chem.* **2022**, *157*, 116727. [CrossRef] [PubMed]
63. Chen, L.; Doeven, E.H.; Wilson, D.J.; Kerr, E.; Hayne, D.J.; Hogan, C.F.; Yang, W.; Pham, T.T.; Francis, P.S. Co-reactant electrogenerated chemiluminescence of iridium(III) complexes containing an acetylacetonate ligand. *ChemElectroChem* **2017**, *4*, 1797–1808. [CrossRef]
64. Guo, W.; Ding, H.; Gu, C.; Liu, Y.; Jiang, X.; Su, B.; Shao, Y. Potential-resolved multicolor electrochemiluminescence for multiplex immunoassay in a single sample. *J. Am. Chem. Soc.* **2018**, *140*, 15904–15915. [CrossRef] [PubMed]
65. Ding, Z.; Quinn, B.M.; Haram, S.K.; Pell, L.E.; Korgel, B.A.; Bard, A.J. Electrochemistry and electrogenerated chemiluminescence from silicon nanocrystal quantum dots. *Science* **2002**, *296*, 1293–1297. [CrossRef]
66. Chen, X.; Liu, Y.; Ma, Q. Recent advances in quantum dot-based electrochemiluminescence sensors. *J. Mater. Chem. C* **2018**, *6*, 942–959. [CrossRef]
67. Xu, Y.; Liu, J.; Gao, C.; Wang, E. Applications of carbon quantum dots in electrochemiluminescence: A mini review. *Electrochem. Commun.* **2014**, *48*, 151–154. [CrossRef]
68. Yang, E.; Zhang, Y.; Shen, Y. Quantum dots for electrochemiluminescence bioanalysis-A review. *Anal. Chim. Acta* **2021**, 339140. [CrossRef]
69. Wang, X.-Y.; Che, Z.-Y.; Bao, N.; Qing, Z.; Ding, S.-N. Recent advances in II-VI quantum dots based-signal strategy of electrochemiluminescence sensor. *Talanta Open* **2022**, *5*, 100088. [CrossRef]

70. Yang, X.; Yu, Y.-Q.; Peng, L.-Z.; Lei, Y.-M.; Chai, Y.-Q.; Yuan, R.; Zhuo, Y. Strong electrochemiluminescence from MOF accelerator enriched quantum dots for enhanced sensing of trace cTnI. *Anal. Chem.* **2018**, *90*, 3995–4002. [CrossRef]
71. Peng, H.; Huang, Z.; Wu, W.; Liu, M.; Huang, K.; Yang, Y.; Deng, H.; Xia, X.; Chen, W. Versatile high-performance electrochemiluminescence ELISA platform based on a gold nanocluster probe. *ACS Appl. Mater. Interfaces* **2019**, *11*, 24812–24819. [CrossRef] [PubMed]
72. Han, S.; Zhao, Y.; Zhang, Z.; Xu, G. Recent advances in electrochemiluminescence and chemiluminescence of metal nanoclusters. *Molecules* **2020**, *25*, 5208. [CrossRef] [PubMed]
73. Nie, Y.; Tao, X.; Zhang, H.; Chai, Y.-q.; Yuan, R. Self-assembly of gold nanoclusters into a metal–organic framework with efficient electrochemiluminescence and their application for sensitive detection of rutin. *Anal. Chem.* **2021**, *93*, 3445–3451. [CrossRef] [PubMed]
74. Jiao, Y.; Hu, R.; Wang, Q.; Fu, F.; Chen, L.; Dong, Y.; Lin, Z. Tune the Fluorescence and Electrochemiluminescence of Graphitic Carbon Nitride Nanosheets by Controlling the Defect States. *Chem. Eur. J.* **2021**, *27*, 10925–10931. [CrossRef] [PubMed]
75. Zou, R.; Teng, X.; Lin, Y.; Lu, C. Graphitic carbon nitride-based nanocomposites electrochemiluminescence systems and their applications in biosensors. *TrAC Trends Anal. Chem.* **2020**, *132*, 116054. [CrossRef]
76. Liu, F.; Du, F.; Yuan, F.; Quan, S.; Guan, Y.; Xu, G. Electrochemiluminescence bioassays based on carbon nitride nanomaterials and 2D transition metal carbides. *Curr. Opin. Electrochem.* **2022**, *34*, 100981. [CrossRef]
77. Ahmad, R.; Tripathy, N.; Khosla, A.; Khan, M.; Mishra, P.; Ansari, W.A.; Syed, M.A.; Hahn, Y.-B. Recent advances in nanostructured graphitic carbon nitride as a sensing material for heavy metal ions. *J. Electrochem. Soc.* **2019**, *167*, 037519. [CrossRef]
78. Wang, Z.; Wei, W.; Shen, Y.; Liu, S.; Zhang, Y. Carbon nitride–based biosensors. In *Biochemical Sensors: Nanomaterial-Based Biosensing and Application in Honor of the 90th Birthday of Prof. Shaojun Dong*; World Scientific: Singapore, 2021; pp. 175–225.
79. Ji, J.; Wen, J.; Shen, Y.; Lv, Y.; Chen, Y.; Liu, S.; Ma, H.; Zhang, Y. Simultaneous noncovalent modification and exfoliation of 2D carbon nitride for enhanced electrochemiluminescent biosensing. *J. Am. Chem. Soc.* **2017**, *139*, 11698–11701. [CrossRef]
80. Huo, X.-L.; Lu, H.-J.; Xu, J.-J.; Zhou, H.; Chen, H.-Y. Recent advances of ratiometric electrochemiluminescence biosensors. *J. Mater. Chem. B* **2019**, *7*, 6469–6475. [CrossRef]
81. Han, Z.; Shu, J.; Liang, X.; Cui, H. Label-free ratiometric electrochemiluminescence aptasensor based on nanographene oxide wrapped titanium dioxide nanoparticles with potential-resolved electrochemiluminescence. *Anal. Chem.* **2019**, *91*, 12260–12267. [CrossRef]
82. Liu, Y.; Sun, Y.; Yang, M. A double-potential ratiometric electrochemiluminescence platform based on gC_3N_4 nanosheets (gC_3N_4 NSs) and graphene quantum dots for Cu^{2+} detection. *Anal. Methods* **2021**, *13*, 903–909. [CrossRef] [PubMed]
83. Fu, X.; Tan, X.; Yuan, R.; Chen, S. A dual-potential electrochemiluminescence ratiometric sensor for sensitive detection of dopamine based on graphene-CdTe quantum dots and self-enhanced Ru(II) complex. *Biosens. Bioelectron.* **2017**, *90*, 61–68. [CrossRef] [PubMed]
84. Zhao, H.-F.; Liang, R.-P.; Wang, J.-W.; Qiu, J.-D. A dual-potential electrochemiluminescence ratiometric approach based on graphene quantum dots and luminol for highly sensitive detection of protein kinase activity. *Chem. Commun.* **2015**, *51*, 12669–12672. [CrossRef] [PubMed]
85. Ye, J.; Zhu, L.; Yan, M.; Zhu, Q.; Lu, Q.; Huang, J.; Cui, H.; Yang, X. Dual-wavelength ratiometric electrochemiluminescence immunosensor for cardiac troponin I detection. *Anal. Chem.* **2018**, *91*, 1524–1531. [CrossRef] [PubMed]
86. Huo, X.-L.; Zhang, N.; Yang, H.; Xu, J.-J.; Chen, H.-Y. Electrochemiluminescence resonance energy transfer system for dual-wavelength ratiometric miRNA detection. *Anal. Chem.* **2018**, *90*, 13723–13728. [CrossRef]
87. Fan, Z.; Yao, B.; Ding, Y.; Xu, D.; Zhao, J.; Zhang, K. Rational engineering the DNA tetrahedrons of dual wavelength ratiometric electrochemiluminescence biosensor for high efficient detection of SARS-CoV-2 RdRp gene by using entropy-driven and bipedal DNA walker amplification strategy. *Chem. Eng. J.* **2022**, *427*, 131686. [CrossRef]
88. Qin, D.; Meng, S.; Wu, Y.; Mo, G.; Jiang, X.; Deng, B. Design of a Dual-Wavelength Ratiometric Electrochemiluminescence Immunosensor for Sensitive Detection of Amyloid-β Protein in Human Serum. *ACS Sustain. Chem. Eng.* **2021**, *9*, 7541–7549. [CrossRef]
89. Wang, P.; Nie, Y.; Tian, Y.; Liang, Z.; Xu, S.; Ma, Q. A whispering gallery mode-based surface enhanced electrochemiluminescence biosensor using biomimetic antireflective nanostructure. *Chem. Eng. J.* **2021**, *426*, 130732. [CrossRef]
90. Li, X.; Du, X. Surface enhanced electrochemiluminescence of the $[Ru(bpy)_3]^{2+}$/tripropylamine system by Au@SiO_2 nanoparticles for highly sensitive and selective detection of dopamine. *Microchem. J.* **2022**, *176*, 107224. [CrossRef]
91. Ding, L.; Xu, S.; Huang, Y.; Yao, Y.; Wang, Y.; Chen, L.; Zeng, Y.; Li, L.; Lin, Z.; Guo, L. Surface-Enhanced Electrochemiluminescence Imaging for Multiplexed Immunoassays of Cancer Markers in Exhaled Breath Condensates. *Anal. Chem.* **2022**, *94*, 7492–7499. [CrossRef]
92. Li, M.-X.; Feng, Q.-M.; Zhou, Z.; Zhao, W.; Xu, J.-J.; Chen, H.-Y. Plasmon-enhanced electrochemiluminescence for nucleic acid detection based on gold nanodendrites. *Anal. Chem.* **2018**, *90*, 1340–1347. [CrossRef] [PubMed]
93. Feng, X.; Han, T.; Xiong, Y.; Wang, S.; Dai, T.; Chen, J.; Zhang, X.; Wang, G. Plasmon-enhanced electrochemiluminescence of silver nanoclusters for microRNA detection. *ACS Sens.* **2019**, *4*, 1633–1640. [CrossRef] [PubMed]
94. Wang, D.; Guo, L.; Huang, R.; Qiu, B.; Lin, Z.; Chen, G. Surface enhanced electrochemiluminescence of $[Ru (bpy)_3]^{2+}$. *Sci. Rep.* **2015**, *5*, 1–7.

95. Wang, D.; Zhou, J.; Guo, L.; Qiu, B.; Lin, Z. A surface-enhanced electrochemiluminescence sensor based on Au-SiO_2 core–shell nanocomposites doped with [Ru (bpy)$_3$]$^{2+}$ for the ultrasensitive detection of prostate-specific antigen in human serum. *Analyst* **2020**, *145*, 132–138. [CrossRef]
96. Kitte, S.A.; Tafese, T.; Xu, C.; Saqib, M.; Li, H.; Jin, Y. Plasmon-enhanced quantum dots electrochemiluminescence aptasensor for selective and sensitive detection of cardiac troponin I. *Talanta* **2021**, *221*, 121674. [CrossRef]
97. Chen, X.; Gui, W.; Ma, Q. Ultrasensitive detection of EGFR gene based on surface plasmon resonance enhanced electrochemiluminescence of CuZnInS quantum dots. *Anal. Chim. Acta* **2018**, *1009*, 73–80. [CrossRef]
98. Zhang, Q.; Liu, Y.; Nie, Y.; Ma, Q.; Zhao, B. Surface plasmon coupling electrochemiluminescence assay based on the use of AuNP@C_3N_4QD@mSiO_2 for the determination of the Shiga toxin-producing Escherichia coli (STEC) gene. *Microchim. Acta* **2019**, *186*, 1–9. [CrossRef]
99. Wang, D.; Guo, L.; Huang, R.; Qiu, B.; Lin, Z.; Chen, G. Surface enhanced electrochemiluminescence for ultrasensitive detection of Hg^{2+}. *Electrochim. Acta* **2014**, *150*, 123–128. [CrossRef]
100. Zhang, Q.; Tian, Y.; Liang, Z.; Wang, Z.; Xu, S.; Ma, Q. DNA-mediated Au–Au dimer-based surface plasmon coupling electrochemiluminescence sensor for BRCA1 gene detection. *Anal. Chem.* **2021**, *93*, 3308–3314. [CrossRef]
101. Zhang, Q.; Zhang, X.; Ma, Q. Recent advances in visual electrochemiluminescence analysis. *J. Anal. Test.* **2020**, *4*, 92–106. [CrossRef]
102. Liu, Y.; Chen, X.; Wang, M.; Ma, Q. A visual electrochemiluminescence resonance energy transfer/surface plasmon coupled electrochemiluminescence nanosensor for Shiga toxin-producing Escherichia coli detection. *Green Chem.* **2018**, *20*, 5520–5527. [CrossRef]
103. Xing, H.; Zhai, Q.; Zhang, X.; Li, J.; Wang, E. Boron nitride quantum dots as efficient coreactant for enhanced electrochemiluminescence of ruthenium (II) tris (2, 2′-bipyridyl). *Anal. Chem.* **2018**, *90*, 2141–2147. [CrossRef] [PubMed]
104. Zhang, Z.; Du, P.; Pu, G.; Wei, L.; Wu, Y.; Guo, J.; Lu, X. Utilization and prospects of electrochemiluminescence for characterization, sensing, imaging and devices. *Mater. Chem. Front.* **2019**, *3*, 2246–2257. [CrossRef]

Article

ImmunoDisk—A Fully Automated Bead-Based Immunoassay Cartridge with All Reagents Pre-Stored

Benita Johannsen [1,*], Desirée Baumgartner [2], Lena Karkossa [1], Nils Paust [1,2], Michal Karpíšek [3,4], Nagihan Bostanci [5], Roland Zengerle [1,2] and Konstantinos Mitsakakis [1,2,*]

1 Hahn-Schickard, Georges-Koehler-Allee 103, 79110 Freiburg, Germany; lena.karkossa@gmx.de (L.K.); nils.paust@hahn-schickard.de (N.P.); roland.zengerle@hahn-schickard.de (R.Z.)
2 Laboratory for MEMS Applications, IMTEK—Department of Microsystems Engineering, University of Freiburg, Georges-Koehler-Allee 103, 79110 Freiburg, Germany; desiree.baumgartner@hahn-schickard.de
3 BioVendor-Laboratorní Medicína a.s., Research & Diagnostic Products Division, Karasek 1767/1, Reckovice, 62100 Brno, Czech Republic; karpisek@biovendor.com
4 Faculty of Pharmacy, Masaryk University, Palackeho trida 1946/1, 61242 Brno, Czech Republic
5 Section of Oral Health and Periodontology, Division of Oral Diseases, Department of Dental Medicine, Karolinska Institutet, 14104 Huddinge, Sweden; nagihan.bostanci@ki.se
* Correspondence: benita.johannsen@hahn-schickard.de (B.J.); konstantinos.mitsakakis@hahn-schickard.de (K.M.); Tel.: +49-761-203-7252 (B.J.); +49-761-203-73252 (K.M.)

Abstract: In this paper, we present the ImmunoDisk, a fully automated sample-to-answer centrifugal microfluidic cartridge, integrating a heterogeneous, wash-free, magnetic- and fluorescent bead-based immunoassay (bound-free phase detection immunoassay/BFPD-IA). The BFPD-IA allows the implementation of a simple fluidic structure, where the assay incubation, bead separation and detection are performed in the same chamber. The system was characterized using a C-reactive protein (CRP) competitive immunoassay. A parametric investigation on air drying of protein-coupled beads for pre-storage at room temperature is presented. The key parameters were buffer composition, drying temperature and duration. A protocol for drying two different types of protein-coupled beads with the same temperature and duration using different drying buffers is presented. The sample-to-answer workflow was demonstrated measuring CRP in 5 µL of human serum, without prior dilution, utilizing only one incubation step, in 20 min turnaround time, in the clinically relevant concentration range of 15–115 mg/L. A reproducibility assessment over three disk batches revealed an average signal coefficient of variation (CV) of 5.8 ± 1.3%. A CRP certified reference material was used for method verification with a concentration CV of 8.6%. Our results encourage future testing of the CRP-ImmunoDisk in clinical studies and its point-of-care implementation in many diagnostic applications.

Keywords: immunoassay; bound-free phase; micro/nanoparticles; point-of-care; centrifugal microfluidics; inflammation; reagent storage

Citation: Johannsen, B.; Baumgartner, D.; Karkossa, L.; Paust, N.; Karpíšek, M.; Bostanci, N.; Zengerle, R.; Mitsakakis, K. ImmunoDisk—A Fully Automated Bead-Based Immunoassay Cartridge with All Reagents Pre-Stored. *Biosensors* **2022**, *12*, 413. https://doi.org/10.3390/bios12060413

Received: 13 May 2022
Accepted: 7 June 2022
Published: 14 June 2022

1. Introduction

Immunoassays can specifically detect protein biomarkers in various sample matrices such as serum, urine or saliva via the binding of an antigen by antibodies [1–3]. They are utilized in many diagnostic applications, ranging from infectious diseases such as malaria, dengue, respiratory tract infections or sepsis to non-communicable diseases such as cardiovascular and autoimmune diseases, periodontal disease, systemic inflammation and many more [1,2,4–9]. Most of these applications benefit from a shift from centralized laboratories towards the point of care (PoC) [2,5,10,11]. Such an approach significantly shortens the time from sample collection to test result, which can be crucial for many of the aforementioned applications, especially when acute and time-critical conditions are involved [12]. The implementation of tests at the PoC determines the following criteria from the end user's perspective, and consequently defines technical requirements: (i) full automation of the analytical workflow so that the test can be performed even by untrained personnel, which requires that the test should have all the necessary (bio)chemical reagents pre-stored and should operate in a sample-to-answer manner; (ii) a rapid turnaround time (TAT), which necessitates a simple and short assay workflow with a low number and complexity of steps; (iii) capability for testing several samples per run (throughput), which requires that the assay reagents occupy as little space as possible on the fully integrated test cartridge; and (iv) optimally, the simultaneous detection (multiplexing) of different biomarkers, which requires the detection method to be compatible with multiplexing configurations.

Centrifugal microfluidic systems are suitable candidates for the shifting of diagnostic practice from centralized to PoC settings. Their closed cartridges can support fully automated analyses, do not require external pumps for fluid handling because they utilize changes in frequency and temperature to run their protocols, and are less prone to bubble clogging [13,14]. Pre-analytic protocols such as blood–plasma separation [15] and pre-treatment of whole saliva [16] can be easily implemented. These are some reasons why many different centrifugal system-based immunoassay solutions have been shown, using different assay methods and with different applications [17–40].

One main structural feature of most of the demonstrated immunoassays based on centrifugal systems is that they use magnetic or fluorescent micro/nanoparticles (in fact, only a few systems do not [18,22,26,28,39]). This is not surprising, as particles have been proven advantageous not only in miniaturized systems. Using magnetic particles as a solid phase shortens the incubation duration [41] and simplifies the handling during different assay steps, for example during washing [42]. The use of fluorescent beads as detection agents simplifies the realization of multiplexing, as a broad range of colors are commercially available [29,30], and they do not require any additional buffers, e.g., to activate and stop in case of an enzymatic reaction, while still retaining the required sensitivity [41,43,44].

The inclusion of micro/nanoparticles coated with proteins (either antibodies or antigens) as assay components requires solutions for their pre-storage directly on the cartridge in order to achieve fully automated sample-to-answer analysis at the PoC [13]. The pre-storage of proteins at room temperature (RT) is in itself a complex topic [45]. The pre-storage of proteins on beads and at RT further increases this complexity, because issues such as unwanted agglutination, resuspension efficiency and stabilization of the beads during the storage procedure are added to the equation. Additionally, as the protein-coupled beads must be stored inside the microfluidic cartridge, the chosen pre-storage process should be compatible with existing, scalable microfluidic cartridge manufacturing technologies. Despite the above challenges, this topic is barely addressed in existing literature. In fact, only a few publications on bead-based centrifugal microfluidic immunoassays report the pre-storage of reagents on the cartridge at RT [24,25,35]. However, none of these contain details on the topic or on the rationale behind the selection of the specific materials and protocols used, which makes it difficult to transfer and adapt their solution to different applications and different proteins. Among these, Lin et al. [24] and Lutz et al. [25] do not implement bead-based ELISA, but lateral flow-based assays with spotted detection lines, which include additional surface treatment steps or the insertion of a membrane into the cartridge, and therefore, their assays are, in their principle, fundamentally different to ours. Zhao et al. [35] automated a standard bead-based ELISA workflow, which, however, requires the pre-storage of four liquid components, includes absorption as a detection method (which does not allow multiplexing, as in the case of fluorescence), and detects ex situ after transferring the resulting liquid to a benchtop equipment. Moreover, all three approaches had to use some method to remove unbound reagents after the incubation in order to reduce the non-specific interactions during incubation and prevent a crosstalk of signal generation during detection. When using washing buffers, this has an impact on reagent consumption and the complexity of the microfluidic design, and although a few wash-free, bead-based configurations have been reported (Schaff et al. [30] and Gao et al. [40]), they implement an additional density medium on a centrifugal cartridge in order to achieve the wash-free operation.

In the context of the state of the art, we present a fully integrated centrifugal microfluidic system, the ImmunoDisk, implementing, for the first time, an immunoassay method that combines a bead-based and wash-free configuration without additional surface treatments or density media, with all reagents pre-stored, and using bound-free phase detection (BFPD). The bound-free phase detection immunoassay (BFPD-IA) is a heterogeneous assay based on ELISA principles and utilizes magnetic and fluorescent beads, the former acting as capture phase and the latter as the detection agent (further information on the BFPD-IA in Johannsen et al. [46]). The simplicity of the BFPD-IA concept allows the realization of assay incubation, bead separation and detection all in a single chamber. This reduces the overall footprint on disk, simplifies the liquid reagent pre-storage and makes the fluidic workflow more robust. Furthermore, the overall workflow of the BFPD-IA on disk and the fluidic structure are significantly simplified compared to the automation of standard ELISA workflow which, next to washing steps, needs to implement enzymatic reactions with the respective additional reagents [23,31,34,35]. Furthermore, in this work, detailed results of a parametric investigation on the pre-storage at RT of antibody-coated magnetic and antigen-coated fluorescent beads are shown for the first time, to the best of our knowledge, with the goal to use one common set of air drying conditions for a simplified disk production. For this, we investigated diverse drying buffer compositions by testing different additives and evaluated drying parameters such as temperature and duration.

The sample-to-answer ImmunoDisk performance in terms of robustness and reproducibility was assessed by detecting C-reactive protein (CRP) in human serum and by using a certified reference material (CRM). CRP was selected for demonstration due to its broad use as an inflammation biomarker [47] to support antibiotic stewardship. The properties of the competitive BFPD-IA on disk allow the direct usage of human serum without any pre-dilution steps in the clinically relevant range of 20 mg/L to 100 mg/L, as proposed by official guidelines in The Netherlands [48] or in the U.K. [49] for supporting clinicians on the decision of antibiotic prescriptions for patients at risk of pneumonia. This makes the ImmunoDisk attractive for the usage at the PoC for the detection of CRP, which is a highly concentrated marker in human serum [4,50,51].

Overall, our parametric study on the pre-storage of protein-coupled beads, in combination with the simplified, wash-free microfluidic integration of the BFPD-IA workflow, can open up new perspectives on the integration of heterogeneous, bead-based immunoassays within (centrifugal) microfluidic systems for user-friendly, rapid, sample-to-answer immunoassay-based diagnostics.

2. Materials and Methods

2.1. Bound-Free Phase Detection Immunoassay

The novel BFPD immunoassay was presented and evaluated in detail in a previous publication by Johannsen et al. [46]. The following is a brief summary of the procedure for the detection of CRP in human serum. The components of the competitive assay are the antibody-coupled magnetic beads as capture agents, the fluorescent beads coupled with the native competitive antigen as detection agents and one assay buffer (without any additional wash buffer). All reagents, together with the human serum sample or standards (spiked antigen concentrations in CRP-free human serum, HyTest, Turku, Finland), are added in a single reaction well (one-step assay) and incubated for 15 min at 37 °C. After this incubation, the signal from the unbound fluorescent beads, correlating to the concentration of CRP in the sample or standards, is measured.

In order to assess the activity of the dried bead-coupled antibodies and antigens after pre-storage and resuspension, we used a negative control of the CRP assay in a microtiter plate, as described in a previous publication [46]. This negative control assay includes both types of coupled beads without the native CRP antigen in the sample. Due to the nature of the BFPD-IA, the negative control facilitates an actual immunoassay reaction which, in the absence of native target analyte, is expected to lead to the maximum binding of the fluorescent bead-coupled competitive CRP antigen to the antibodies on the magnetic beads. Because the detection occurs in the (remaining) bound-free phase of the fluorescent bead suspension, such a negative control BFPD-IA gives the lowest signal, since no competition with any native antigen in the sample takes place.

2.2. Preparation of Magnetic Beads for the CRP Assay

The coupling of anti-human CRP antibodies (A80-125A, Fortis Life Sciences (Bethyl), Waltham, MA, USA) onto the tosyl-activated surface of the magnetic beads was described in detail in a previous publication by Johannsen et al. [46]. In short, the anti-human CRP antibodies were coupled onto the tosyl-activated surfaces of magnetic beads with a diameter of 2.8 μm (Dynabeads, M-280, Thermo Fisher Scientific, Waltham, MA, USA) overnight (18 h) at 37 °C under rotation. This was followed by a blocking step with 0.5% bovine serum albumin (BSA) (Carl Roth, Karlsruhe, Germany) for 2 h and a deactivation step with 50 mM ethanolamine (Carl Roth, Karlsruhe, Germany) for 1 h. After being washed twice, the beads were stored at 2.0% solid in a storage buffer (PBS (Thermo Fisher Scientific, Waltham, MA, USA), 0.1% BSA, 0.03% Synperonic P84 (Sigma-Aldrich, St. Louis, MO, USA), 0.05% sodium azide (Carl Roth, Karlsruhe, Germany)) at 4 °C until further use.

2.3. Preparation of Fluorescent Beads for the CRP Assay

The coupling of native CRP protein (C7907-26, 95–98%, highly purified, United State Biological, Salem, USA) onto the carboxyl-activated surfaces of the fluorescent beads (F8810, red (excitation 580 nm/emission 605 nm), 0.2 μm, Thermo Fisher Scientific, Waltham, MA, USA) was described in detail in a previous publication by Johannsen et al. [46]. In short, the carboxylated surface was activated with 25 mM EDC (1-ethyl-3-(3-dimethylaminopropyl)carbodiimide hydrochloride) (Thermo Fisher Scientific, Waltham, MA, USA) and 25 mM NHS (N-hydroxysuccinimide) (Thermo Fisher Scientific, Waltham, MA, USA). After activation, the CRP protein was coupled onto the surface in the presence of a 25 mM MES (2-(N-morpholino)ethanesulfonic acid) buffer with a pH of 6.1. A post-saturation step with 1.0% BSA and hydrolyzation of the remaining active groups with ethanolamine followed. After two washing steps, the beads were stored at 2.0% solid at 4 °C.

2.4. Parametric Investigation of Pre-Storage of Protein-Coupled Beads

For the evaluation of the drying buffer as part of the parametric study, 5 μL of drying buffer containing either the magnetic or the fluorescent beads was dried in a microtiter plate (96 wells, polystyrene, non-binding, Greiner Bio-One, Frickenhausen, Austria) in an incubator (INCU-Line IL 23, VWR International, Radnor, PA, USA) at 37 °C for 17 h and 6 h for the magnetic and fluorescent beads, respectively. They were stored in the dark in a container with silica beads. The beads were resuspended for 5 min at 750 rpm (BioShake iQ, QInstruments, Jena, Germany) before the CRP assay was conducted (see Section 2.1). The drying buffer was composed of different combinations of single or multiple additives.

2.5. Drying Buffer for the Magnetic Beads

The following additives to PBS were used (in different combinations) to test different drying buffer compositions for drying the anti-CRP antibody-coupled magnetic beads: (1) trehalose (Sigma-Aldrich, St. Louis, MO, USA), (2) sucrose (Sigma-Aldrich, St. Louis, MO, USA), (3) BSA, (4) Polyethylene glycol 1000 (PEG1000) (Merck KGaA, Darmstadt, Germany), (5) Tween80 (Sigma-Aldrich, St. Louis, MO, USA) and (6) CHAPS (Sigma-Aldrich, St. Louis, MO, USA). Additionally, two buffers from the companies Merck KGaA (Darmstadt, Germany) and GSK (Brentford, UK), which they use for the lyophilization of their proteins, were tested. The detailed composition of the additives in these buffers is given by Mensink et al. [45] (we dissolved them in PBS for comparison reasons). Two additional components that were not part of the overall parametric investigation, L-histidine (Sigma-

Aldrich, St. Louis, MO, USA) and sodium phosphate dibasic heptahydrate (Sigma-Aldrich, St. Louis, MO, USA), were utilized for these buffers.

2.6. Drying Buffer for the Fluorescent Beads

The following additives to PBS were used (in different combinations) to test different drying buffer compositions for drying the CRP antigen-coupled fluorescent beads: sucrose, trehalose, PEG1000 and Tween80.

2.7. ImmunoDisk Cartridge Fabrication

The centrifugal microfluidic disks were designed using SOLIDWORKS 19 (Dassault Systèmes, Vélizy-Villacoublay, France) and simulated utilizing a network simulation processed with MATLAB Simulink Simscape R2016a (The Mathworks, Natick, MA, USA) that was improved for the development of centrifugal microfluidic disks, as described in detail by Schwarz et al. [52]. The disks were fabricated by means of thermoforming thin polycarbonate (PC) foils (Makrofol, thickness: 250 μm, Covestro AG, Leverkusen, Germany), as described by Focke et al. [53]. Instead of an elastomeric mold, as used by Focke et al., for the production of the cartridges in our work, a metal master tool was milled (EVO, KERN Microtechnik GmbH, Eschenlohe, Germany) and used for automatic production (Rohrer AG, Möhlin, Switzerland). The PC foil disks were punched to create an inner hole (10 mm diameter) and outer rim (130 mm diameter) that allowed exact placement of the disk on the processing device, the LabDisk Player 1 (DIALUNOX GmbH, Stockach, Germany). Magnetic and fluorescent beads were pre-stored on the cartridge in one step. The magnetic particles with anti-human CRP antibodies on their surfaces were pipetted from their stock, and the buffer was exchanged with the drying buffer (PBS, 10% (w/v) trehalose and 50% (w/v) PEG1000) to give a final bead concentration of 20 μg/μL. The fluorescent beads with native CRP antigen on their surfaces were diluted directly in their drying buffer containing PBS with 10% (w/v) trehalose and 10% (w/v) sucrose to give a final bead concentration of 3.8 μg/μL. Then, 5 μL of each of the magnetic and the fluorescent bead solutions was pipetted into dedicated storage chambers on the disk and then dried in an incubator at 39 °C for 1 h. The disks were stored at RT in a container with silica beads (SGT002, Silica Gel Shop, Haaksbergen, The Netherlands) before sealing. The assay buffer (dilution buffer, BioVendor, Brno, Czech Republic) was stored in stickpacks (described in detail by van Oordt et al. [54]) that are designed to open at a specific frequency on the disk. The disk was sealed with a pressure-sensitive adhesive film (9795R, 3M, Saint Paul, MN, USA). The vents and inlets were opened with a scalpel. The disks with all reagents pre-stored were kept in a container with silica beads at RT and protected from light.

3. Results and Discussion

3.1. ImmunoDisk—Description of Fluidic Workflow

A centrifugal microfluidic disk, the ImmunoDisk, was developed for the automation of the BFPD-IA workflow with all reagents pre-stored (Figure 1A,B). The assay buffer (110 μL) is stored in stickpacks, as previously described in detail [54]. The magnetic and fluorescent beads are stored in dedicated storage chambers via air drying. Details on the pre-storage process can be found in Section 3.5. An overview of the fluidic structure and the workflow is shown in Figure 1A. Three such fluidic structures can fit in a full disk, which allows the testing of three samples in parallel.

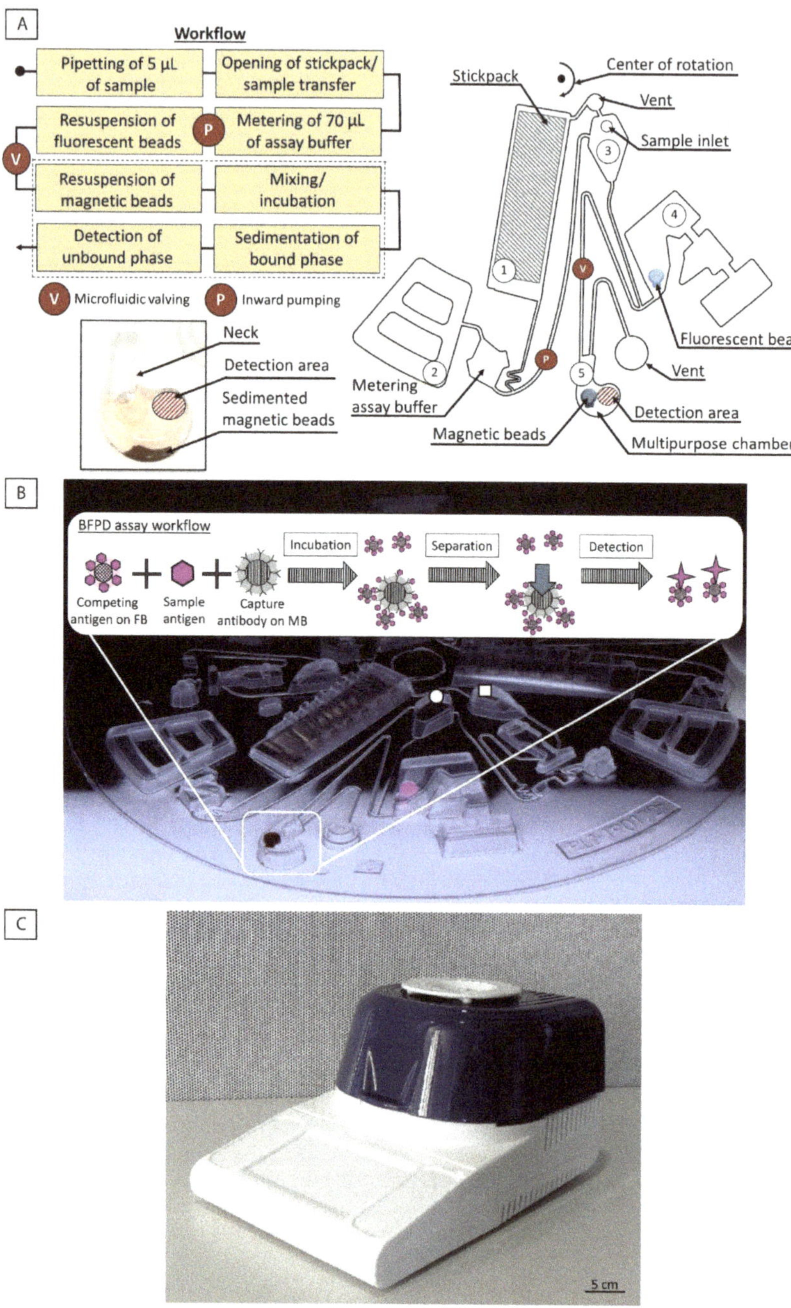

Figure 1. (**A**) Overview of the fluidic structure and workflow of the ImmunoDisk for the automation of the BFPD-IA. #1: stickpack chamber; #2: pneumatic valve chamber; #3: inlet chamber; #4: valve

chamber; #5: multipurpose chamber. The workflow consists of only one inward pumping and one valving step. A specially designed multipurpose chamber (bottom left, #5) is used for storage of the magnetic beads, incubation (mixing), separation (sedimentation) and detection. The detailed fluidic workflow and the function of chambers #1–5 are described in Section 3.1. The parts of the workflow that are surrounded by a dashed line, all take place in the multipurpose chamber. (**B**) ImmunoDisk with stored magnetic (brown) and fluorescent beads (pink), shown here in liquid form before air drying, for better visibility. The assay buffer is stored in one stickpack that releases the buffer upon centrifugation. The schematic inset shows how the whole BFPD-IA workflow, including incubation, separation of the magnetic beads (MB) and detection of the unbound fluorescent beads (FB), takes place in the multipurpose chamber. Three complete structures fit on one cartridge. The drawn circle indicates the inlet for serum that was used in this study. The drawn square indicates the inlet for a potential whole blood sample, followed by a microfluidic module for plasma separation. This was not used in this study but was included in the design in order to assess the maximum amount of space consumed overall and for future usage. (**C**) Image of the processing PoC device (functional model), the LabDisk Player 1. The device was developed as a PCR cycler and can achieve different temperatures using air heating and is used to run centrifugal microfluidic disks. The device also contains two fluorescence detectors with four different colors in total.

The disk is mounted into the processing device (LabDisk Player 1, see Figure 1C). Then, 5 µL of the sample (human serum) is pipetted into the inlet and the fluidic and temperature protocol is started (Supplementary File S1). The assay buffer in chamber #1 is released by the opening of the seal of the stickpack at a high frequency (80 Hz). Simultaneously, the sample flows from the inlet to chamber #4 (without yet resuspending the pre-stored fluorescent beads as the filling level is not high enough). The assay buffer then flows into a metering chamber to measure 70 µL out of the total 110 µL. Residual liquid is transported into the overflow chamber #2 and additionally loads the pneumatic pumping structure [55], which is activated by reducing the frequency to 7 Hz. The functionality of pneumatic valves and other pneumatic unit operations in centrifugal microfluidics is described in detail by Hess et al. [56]. The 70 µL of buffer is then pneumatically pumped radially inwards into the inlet chamber #3. With an increase in the frequency (40 Hz), the buffer is transferred to chamber #4, containing the stored fluorescent beads and the sample, which was transferred previously. This chamber is not vented, and the presence of liquid loads the valve. The fluorescent beads are resuspended in seconds upon mixing with the sample and the assay buffer. The valve is activated again by reducing the frequency (14 Hz) and by increasing the temperature to 37 °C, which is also the incubation temperature for the immunoassay. The liquid is then transferred into the multipurpose chamber #5, resuspending the magnetic beads. Batch-mode [57] mixing during incubation is started. No external magnets are required to achieve a sufficient mixing result. After incubation, the frequency is increased (30 Hz for 30 s, 40 Hz for 10 s) to sediment the bound-phase and separate it from the bound-free phase (Figure 1A and a photo of the sedimented bound-phase can be found in Supplementary File S2). The disk is stopped so that the multipurpose chamber is located above the detector for the fluorescence signal readout in the bound-free phase. The detection area is well-defined and spatially separated from the sedimented bound phase to prevent any interference with the signal.

3.2. Fluidic Characterization

Some variance in the entire fluidic system, and consequently the measurement result, may in principle derive from volume variances when (i) pipetting the sample into the inlet or pipetting the liquid solutions of fluorescent and magnetic beads prior to drying, (ii) metering the assay buffer, (iii) pumping the assay buffer radially inwards into chamber #3 and (iv) valving and transferring the mixture (sample, assay buffer and resuspended fluorescent beads) from chamber #4 into the multipurpose chamber #5.

The performance of the metering process and the volume transfer of the pneumatic pumping and the pneumatic valving steps were evaluated by pipetting 5 µL of assay buffer into the sample inlet and 110 µL of assay buffer into the stickpack chamber. The fluidic protocol was started and the volume that was transferred into the incubation chamber was measured (N = 6) with an average of 78.3 ± 0.5 µL. This corresponds to a deviation of 4% from the targeted 75 µL and shows a high reproducibility, with a coefficient of variation (CV) below 1%. To assess the reproducibility of the resuspension of the fluorescent beads and the fluidic valving, fluorescent beads were dried in their assigned storage chamber #4 before 75 µL of buffer was pipetted into the inlet to ensure a controlled volume. The buffer was transferred to chamber #4 containing the fluorescent beads, which were resuspended, mixed and transferred to the multipurpose chamber. After batch-mode mixing for 200 s, the fluorescence signal was measured (N = 9). The average signal of 1132 ± 30 RFU showed a CV of only 2.7%. Part of these (anyway small) variances may be attributed to the manual pipetting and variances of the detector. We expect even lower variances for automatically produced disks in the future. The combination of pneumatic and temperature-induced overpressure during valving ensures that all of the resuspended fluorescent beads are transferred to the multipurpose chamber. All these results demonstrate low variation deriving from the microfluidic operations and show that the disk and its fluidic protocol are well-suited for performing the BFPD-IA.

3.3. Multipurpose Chamber

The multipurpose chamber was designed with some special features to facilitate the bead and liquid handling and to optimize the various assay steps, all in a single chamber (Figure 1A). (i) A radially inward 'neck' was introduced to the chamber in order to stabilize the meniscus that forms when the disk stops below the detector for the readout step. The assay buffer used for the detection of CRP in human serum has a contact angle of 74° (measured with Physica MCR101, Anton Paar, Graz, Austria) on the polycarbonate disk material, which makes it wetting (as its contact angle is <90° [58]). (ii) The multipurpose chamber incorporates a small second chamber, which is used for the storage of the magnetic beads (storage chamber, Figure 2C). Notably, the chamber has been positioned a distance away from the planar detection area in order to not interfere with it (Figure 1A). (iii) The radially outward rounding facilitates the collection of the sedimented magnetic beads (Supplementary File S2) and prevents them from interfering with the detection, which is realized within the same chamber.

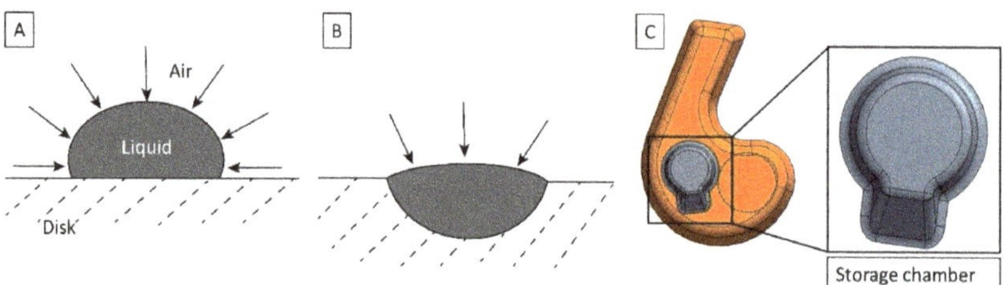

Figure 2. (**A**) Droplet for drying, positioned on the disk planar surface without a storage chamber. The air–liquid interface is defined by the properties of the drying buffer and its contact angle on the cartridge surface. (**B**) Droplet for drying in a dedicated storage chamber on the disk. The size of the air–liquid interface is reduced. (**C**) Multipurpose chamber (orange) with storage chamber (gray) and detection area (black circle). The storage chamber has a volume of 5 µL and includes a ramp that supports the flow of beads radially outwards after resuspension.

A key element of the BFPD-IA method is the detection in the unbound phase, which consists of only the fluorescent beads, after the bound phase (the complexes of magnetic beads with fluorescent beads and/or CRP antigen from the sample) has been removed. Due to its special features, and in combination with the appropriate centrifugal protocols, the multipurpose chamber allows this key element of the BFPD-IA to be transferred to the disk cartridge by integrating the single-step assay incubation (including mixing), the separation of the bound phase (sedimentation) and the detection in the same chamber. This is enabled by the fact that, based on the design of the assay itself, the unbound fluorescent beads are not sedimented, while the bound phase is. The distance traveled by a bead at specific frequencies can be estimated using Stokes' law modified for centrifugal forces [59,60] (equation in Supplementary File S3, where it is shown that the bead radius and density are the key parameters). The extreme case scenario for the sedimentation calculation would consist of a single magnetic bead (nothing bound on its surface) at the most radially inwards position in the multipurpose chamber, and thus traveling the maximal distance of 0.5 cm to a radial position of 6.2 cm. The fluidic protocol consists of two sedimentation steps. The first step sediments with 30 Hz for 30 s and the second step with 40 Hz for 10 s. Using the equation in Supplementary File S3 to calculate the distance traveled gave the final position of a single magnetic bead at 6.2 cm. We applied both sedimentation steps to a single, unbound fluorescent bead, and we calculated that it would move only a few micrometers. Thus, we can be sure that the fluorescent beads are not subjected to sedimentation when using our protocol. Overall, the capability to perform our assay and readout in a single multipurpose chamber has significant positive consequences, as it (i) saves time, (ii) reduces possible errors (no bead loss, metering is omitted, no liquid transfer) and (iii) reduces the occupied space, enabling more structures to fit on one disk. This can subsequently increase the throughput and reduce the manufacturing costs.

3.4. Storage Chamber on the Disk

There can be significant denaturation of proteins at the air–liquid interface during the drying for pre-storage [61]. The air–liquid interface can be reduced by placing the bead-containing solution into a geometrically confined area, rather than simply on the planar surface of the plastic cartridge (Figure 2), thereby reducing the risk of protein denaturation. We have also observed that a droplet on the cartridge surface can become unstable during handling. This leads to the liquid spreading over the surface and a considerable increase in the air–liquid interface. Therefore, the ImmunoDisk consists of two storage chambers (close-up of the design in Figure 2C) that are used to pre-store the fluorescent and magnetic beads directly on the cartridge. Such storage chambers simplify the pipetting and handling during production and allow a more robust fabrication of the complete cartridge (also advantageous for transport) because the air–liquid interface is more stable in the storage chamber than on the planar surface. Furthermore, before we started using the storage chamber, when we simply pipetted the bead solution onto the planar cartridge surface, pellets occasionally broke or became loose during handling after drying. This did not occur after we started using the storage chamber. We also observed better resuspension behavior of the magnetic bead pellet when the storage chamber was used, because of a ramp that supports the flow of beads radially outwards during resuspension (Figure 2C).

3.5. Study on Pre-Storage of Protein-Coupled Beads

Reagent pre-storage on a disposable microfluidic cartridge is essential in order to provide full automation. It reduces the number of hands-on steps and allows easy transportation of the disposable cartridges. Storage of proteins, especially at RT, is challenging. Factors such as heat, shear forces and contact with interfaces can result in a loss of protein functionality due to denaturation [45,61]. There are some theories on how proteins can be protected against denaturation during storage, which can be summarized in two categories: either the mobility of a protein (vitrification theory) or reactions with its surroundings (water replacement theory) must be restricted [45]. Lyophilization or freeze-drying is often used to store proteins (alone, i.e., without beads) due to the good stabilization achieved with this method, even though it is costly and technologically complex [61]. Concerning beads alone, we have demonstrated protocols for pre-storage in the case of magnetic beads for nucleic acid extraction and purification [62–64]. However, the BFPD-IA that is integrated within the ImmunoDisk involves protein-coupled magnetic and fluorescent beads. This results in specific requirements and also restrictions for the selection of the pre-storage method. As the fluorescent beads that were used in this work cannot withstand freezing, according to the manufacturer [65], the lyophilization approach was not an option. Storing the protein-coupled beads in liquid buffer was also excluded, in order to avoid cold chain requirements during transport. Therefore, the air drying method was chosen.

From a scalable manufacturing perspective, it was also judged that air drying would be a simpler workflow compared to lyophilization (i.e., the creation of a pick-and-placeable lyopellet), because the former can be conducted with dispensing robotics and large ovens, while the latter would require a pick-and-place robot and strict environmental humidity conditions.

The requirements for the pre-storage on the ImmunoDisk (and also for bead-based PoC immunoassays in general) were (i) maintaining the functionality of the proteins, (ii) the fast and complete resuspension of the beads and (iii) the prevention of aggregate formation. Additional manufacturing-related requirements demand that there should be (iv) stable bead pellets and, ideally, (v) no additional treatment of the (cartridge) surfaces.

The parameters that may influence the storage result are the drying buffer (including protein stabilizing additives), the drying temperature and the drying duration next to the fixed parameters of bead type, proteins and volumes. Thus, we conducted a detailed experimental investigation of these parameters. In the first step, we tested a non-exhaustive list of candidate drying buffers, temperatures and durations and evaluated each of them by means of assay performance and pellet resuspension. We also took into account that the final drying conditions should be compatible with both bead types (magnetic and fluorescent), as well as both biomolecule types coupled on the beads' surface (anti-human CRP antibodies and native CRP antigens, respectively).

This part of our work should serve as an orientation for the reader in the field of pre-storage of protein-coupled beads. We make no claim of a complete investigation of the drying parameters and conditions, nor was the goal to examine the impact of these conditions on the structure/function of the proteins and beads.

3.6. Parametric Study on Drying Buffers

We started our study by investigating suitable drying buffers first for the antibody-coupled magnetic beads (MB), while keeping the antigen-coated fluorescent beads (FB) in solution, and then for the antigen-coupled fluorescent beads, while keeping the antibody-coated magnetic beads in solution. We used the BFPD-IA, where all reagents were in the liquid state, as a reference. The signal intensity resulting from each assay performed with a specific drying buffer was subtracted from the reference and this ΔIntensity was used to assess the functionality of the reagents after drying (as ΔIntensity shows the difference from the reference, it should be as low as possible). A higher CV could indicate the formation of aggregates or unreproducible resuspension properties and therefore, the CV should be comparable to the CV of the reference assays. Images of dried beads were taken after drying with the attempt to find a correlation between the visual properties of the pellet and the post-rehydration assay performance, but we did not find any. Therefore, we considered only the assay performance itself as an evaluation criterion for the drying process. These evaluation experiments were first performed on a microtiter plate, and the best-performing drying buffers were then tested on the cartridge.

Representative additives were selected from classes of reagents that have been reported to support the stabilization of proteins, such as sugars, polymers and detergents [45,61], or to counteract possible adsorption effects, such as BSA [66].

Different drying buffer combinations were tested for the air drying of MB, and the most representative results are shown in Figure 3A. We started by using PBS, which is used as the base for the drying buffers, alone without additives (listed as 'none'). We observed an (undesirable) high ΔIntensity signal. This indicates that either the beads could not be resuspended completely or the proteins on the surface of the beads lost their binding affinity.

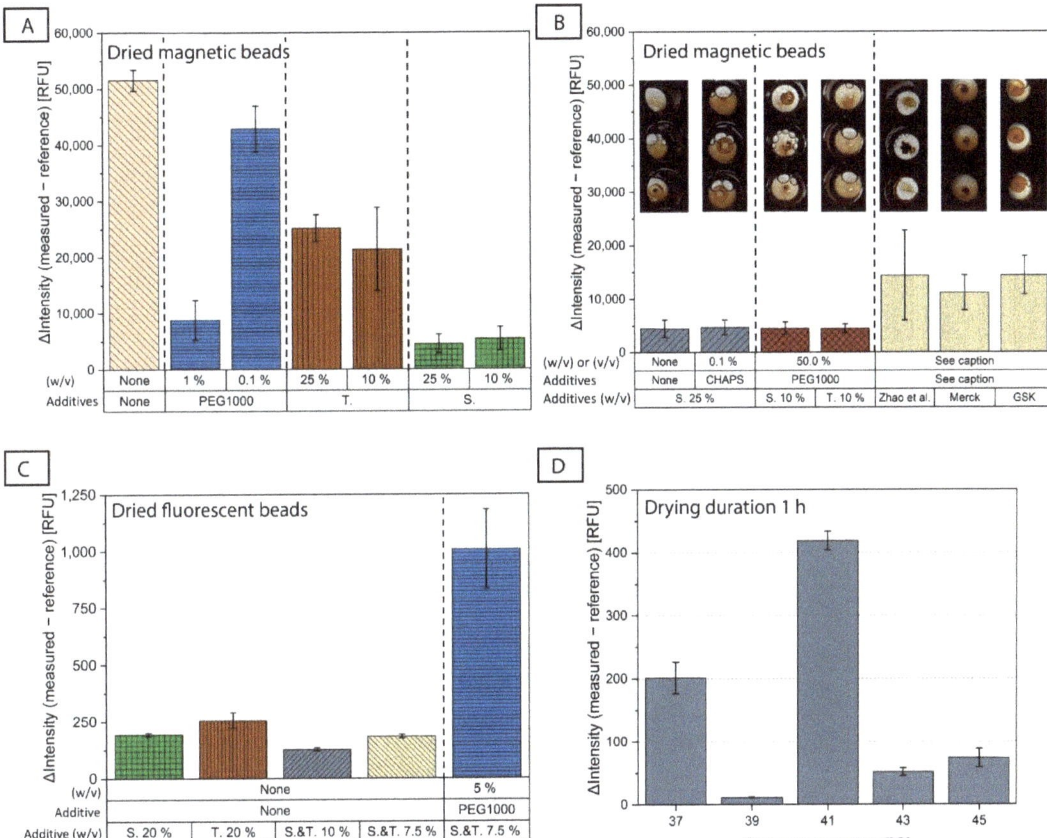

Figure 3. (**A**) Comparison of the assay behavior after air drying of magnetic beads (keeping fluorescent beads in solution) with some representative drying buffers based on PBS and containing only one additive. The additives tested were trehalose (T.), sucrose (S.) and PEG1000, as well as only PBS ('none'). ΔIntensity evaluates the assay performance. N = 3, except S. 25%, which already includes all measurements conducted during the course of this study, for which N = 10. (**B**) The four best-performing drying buffers for magnetic beads from this parametric study were measured again on a second day using more repetitions to ensure reproducible results with a total of N = 10. To show different resuspension behavior, the best-performing buffers are compared to the buffer published by Zhao et al. [35] and buffers from the companies Merck and GSK that were developed for lyophilization of proteins (N = 3; their compositions are given in [45]). The figures above the bar diagrams show the differences in the resuspension of the dried magnetic beads (for more repetitions than N = 3, representative figures were chosen). (**C**) Comparison of the assay behavior after air drying of fluorescent beads (keeping the magnetic beads in solution) with some representative drying buffers (N = 3). (**D**) An overview of the assay performance with different drying temperatures (37 °C (N = 3), 39 °C (N = 4), 41 °C (N = 2), 43 °C (N = 5) and 45 °C (N = 2)) and a common drying duration of 1 h using the chosen drying buffers for magnetic beads (trehalose 10% (*w*/*v*), PEG1000 50% (*w*/*v*) in PBS) and fluorescent beads (trehalose and sucrose 10% (*w*/*v*) in PBS) on the disk. Colors of the bar diagrams are included for better visualization and are not intended for cross-correlation between the figures (**A**) to (**D**).

Two different disaccharides were tested, trehalose and sucrose. Disaccharides are non-reducing sugars and, in contrast to reducing sugars, do not denature proteins with a Maillard reaction [61]. It has also been shown that they can stabilize proteins during storage [45]. For the MB, the ΔIntensity was clearly lower when adding trehalose and sucrose at different concentrations than when using PBS alone (Figure 3A), indicating the positive effect that both disaccharides had on the preservation of the proteins' functionality, with sucrose appearing to perform better. We did not observe any major difference in the ΔIntensity between 10% and 25% (w/v) for either of the two disaccharides.

We also examined a representative polymer additive, PEG1000. PEG as a protein stabilizer has been controversially discussed in the literature and is often said to be used with caution, as its effect is protein-dependent [61,67,68]. While some of its properties, such as the good hydration of proteins in PEG-containing liquids and its ability to adsorb to the hydrophobic core of proteins, indicate good stabilizing potential [69,70], it has also been shown that its usage reduces protein transition temperatures, above which the structure of a protein changes [68]. Experimentally, we saw that when using PEG1000 alone at a concentration of 1% (w/v), we had quite low ΔIntensity values, almost as low as in the case of sucrose (Figure 3A), which implies that it can stabilize the bead-coupled antibodies even at this low concentration. In contrast, lower PEG1000 concentrations of the order of 0.1% had a sub-optimal performance. However, we observed that 1% (w/v) PEG1000 combined with 10% (w/v) trehalose or 10% (w/v) sucrose was not sufficient to improve the results, which were ~3× worse when using the sugars alone (Supplementary File S4A). When we increased the PEG1000 concentration to 25% (w/v) and combined it with 10% (w/v) sucrose or 10% (w/v) trehalose, the ΔIntensity was drastically reduced, although with some large error bars for the case of 10% (w/v) sucrose/25% (w/v) PEG1000 (Supplementary File S4A), which was solved when further increasing the PEG1000 concentration to 50% (w/v) (Supplementary File S4A and Figure 3B). This could lead to the conclusion that if PEG1000 is used as a second additive alongside a disaccharide, its concentration needs to be high for it to have an impact. Another important observation is that although the sugars performed differently when used alone (e.g., 10% (w/v) sucrose vs. 10% (w/v) trehalose, Figure 3A), this difference was no longer observed when PEG1000 was added at a concentration of 50% (Figure 3B, red columns).

Three further additives were combined, in different concentrations, with both 25% (w/v) trehalose and 25% (w/v) sucrose: BSA, which might help to reduce non-specific binding to surfaces during drying [66], and the detergents Tween80 and CHAPS (in powder form at RT), which could prevent the adsorption and aggregation of proteins [61]. According to our results, the addition of BSA did not have a positive effect (Supplementary File S4B,C). We also observed different behavior between Tween80 and CHAPS (Supplementary File S4B,C), which shows that the type of detergent has an influence. Neither improved the results when added to sucrose, but adding 2% (v/v) Tween80 to trehalose reduced the ΔIntensity substantially, although the CV was increased.

Figure 3B shows a comparison of the four buffers which performed best for the drying of MB in this parametric study. We found that 25% (w/v) sucrose, 25% (w/v) sucrose with 0.1% (w/v) CHAPS, 10% (w/v) sucrose with 50% (w/v) PEG1000, and 10% (w/v) trehalose with 50% (w/v) PEG1000 had similar ΔIntensity results. As mentioned earlier, the selection of the drying buffer should be based not only on the degree of stabilization of protein function (assessed through the assay performance), but also on the quality of the resuspension of the dried magnetic bead pellet. The corresponding photo of the well after resuspension of the dried MB is shown in Figure 3B to emphasize the different resuspension behavior depending on the drying buffer used, even though all four buffers showed similar ΔIntensity results. This could indicate that the performance criteria of measured ΔIntensity and its CV represent the combination of the resuspension behavior of the pellet, the protein stabilization properties of the drying buffer and the formation of bead-aggregates after resuspension of the pellet.

We also tested commercial buffers from the companies Merck and GSK [45] that were developed for lyophilization of proteins, as well as the buffer by Zhao et al. [35]. However, these did not offer any significant improvement compared to the drying buffers from our study, neither in terms of assay performance (indicated by the high ΔIntensity), nor in terms of bead resuspension (last three inset photos in Figure 3B).

After the screening of different drying buffers in the microtiter plate, the best-performing buffers for the drying of protein-coupled magnetic beads were tested directly on the ImmunoDisk cartridge (exemplary results in Supplementary File S5). The final buffer that was chosen for use in further experiments was 10% (w/v) trehalose with 50% (w/v) PEG1000 in PBS. It showed superior resuspension quality and reproducibility in the cartridge, as was the case in the microtiter plate (Figure 3B).

The parametric study continued with the investigation of drying buffers for the antigen-coupled fluorescent beads, while keeping the magnetic beads in solution. The parametric study for the drying of native CRP-coated fluorescent beads included fewer candidate drying buffers than in the tests for the magnetic beads, as we utilized findings from the latter study. We did not observe as big a difference between trehalose and sucrose (Figure 3C). While PEG1000 showed promising results for the CRP-antibody coupled magnetic beads, it resulted in an increase in the ΔIntensity after drying of the antigen-coupled fluorescent beads (Figure 3C and Supplementary File S6). The CRP native antigen that is coupled onto the fluorescent bead surfaces is mainly hydrophobic [71], and it has been shown that hydrophobic proteins can be destabilized by PEG [68]. These findings demonstrate that there is no universal solution for additives in drying buffers for different types of proteins (antibody vs. CRP antigen), and that candidate buffers should be tested for each protein individually. The addition of Tween80 at various concentrations also did not improve the results (Supplementary File S6).

Due to the above, we decided to continue with the drying buffer for FB containing only sugars. Figure 3C shows that 10% (w/v) trehalose combined with 10% (w/v) sucrose in PBS gave the lowest ΔIntensity value, and therefore, this drying buffer was selected for the fluorescent beads.

The fact that different drying buffers had to be selected for the antibody-coated magnetic beads and the antigen-coated fluorescent beads might be due to the different nature of the proteins, which we also observed with the different impact of PEG1000 on the drying results. The drying buffer for the CRP antibody-coupled magnetic beads also showed promising results when used for magnetic beads that were coupled with matrix metalloproteinase 9 (MMP-9) antibodies [46] measured on disk (Supplementary File S7). The fact that the same magnetic bead-drying buffer was used successfully with two different antibodies could be a preliminary indicator that it can be used for a range of different antibodies. This should be investigated further in the future.

Our experimental observations indicate that for different types of proteins, it is necessary to test different additives in different concentrations, and that there is no universal solution for all proteins.

3.7. Investigation of Pre-Storage Temperature and Duration Conditions

Having defined the biochemistry of the drying buffers for the magnetic and fluorescent beads, the next step in our parametric study was to define the drying conditions. Unlike the buffers, which do not have to be the same, the duration and temperature must be the same for both drying buffers, as the two bead solutions must be dried simultaneously and in the same cartridge.

We used a design of experiments (DoE) approach with Minitab (Minitab GmbH, Munich, Germany) to evaluate the influence of temperature and duration on the drying of the protein-coupled beads. The DoE showed that the combination of drying temperature and duration has the strongest influence on the result, along with the drying duration alone. The drying temperature by itself has no significant influence on the result. It should be kept in mind that the DoE tested low and high temperatures of 37 °C and 45 °C and that higher temperatures can have a strong influence, for example, if the proteins are denatured. For more information on the DoE please see Supplementary File S8.

The drying duration was set to 1 h, which is favorable from a disk manufacturing perspective. A screening of different drying temperatures, shown in Figure 3D, was conducted to find the best-performing temperature–duration combination. The drying was carried out directly on the ImmunoDisk (in the storage chambers, Figure 1A) and not in the microtiter plate wells. We observed differences in the ΔIntensity and the signal CVs for different drying temperatures (Figure 3D). The negative control that performed best, showing a low CV (3.3%) and a low ΔIntensity (11.3 RFU), was dried at 39 °C. All other temperatures led to either higher CVs (3.4–20.3%) or a higher ΔIntensity. Thus, the final selected drying temperature and duration were 39 °C and 1 h, respectively. Some additional screening with longer drying durations at different temperatures was conducted, which ensured that the drying at 39 °C for 1 h indeed shows superior performance compared to longer drying durations (overview of the data in Supplementary File S9). Using the drying conditions of 39 °C for 1 h, the resuspension was completed in a few seconds for the fluorescent beads and in less than 90 s for the magnetic beads.

Having specified the drying buffers and conditions, we briefly explored the short-term stability (two weeks) of the dried beads in a disk cartridge at RT. We first used dried magnetic beads and liquid fluorescent beads. We compared the average results on day 1 and day 16 using the negative control. In this configuration, we observed that the signal on day 16 after drying differed by 1.7% compared to that on day 1. In a subsequent set of experiments, we used dried magnetic beads as well as dried fluorescent beads. As above, we used the negative control. The negative control was measured after 1, 4, 11 and 15 days of storage at room temperature. In this case, the average signal on day 15 after drying differed by 4.7% compared to day 1. The measurements over the four different days after storage show a variation of 6.3% (N = 7). These are promising preliminary results and the deviations are within the inter-assay variation. The storage stability over time needs to be evaluated further with longer storage durations and over all concentrations covered by the calibration curve, but this would go beyond the scope of this work.

3.8. Sample-to-Answer CRP Detection on the ImmunoDisk

In a final step, the results from the developments in the areas of reagent pre-storage and assay transfer to microfluidics were demonstrated in a sample-to-answer configuration on the disk with all reagents stored at RT. CRP was used as a target antigen and CRP-free human serum as a complex matrix.

A calibration curve was obtained by measuring CRP-spiked human serum standards using three batches of disks on three different days (Figure 4A). This allowed us to perform a reproducibility assessment, which showed an average inter-assay signal CV of 5.8 ± 1.3%. The variation was below 10% and included the influence of the pre-storage of the protein-coupled beads at RT, fluidic variations and production variation. This shows a high reproducibility of the BFPD measurement for CRP in the range between 15 mg/L and 115 mg/L, which includes the clinically relevant cut-off values [51,72]. The limit of detection (LOD) was calculated as 18.2 mg/L. Although some central laboratories use 5 mg/L as a low cut-off value of CRP for general inflammation conditions, clinical studies focusing on specific health conditions and especially respiratory tract infections use 'zones' of CRP concentrations for supporting the decision of prescribing antibiotics or not. In particular, Prins et al. [51] and Jakobsen et al. [72] report a single cut-off value of 50 mg/L, and Alcoba et al. [73] report 80 mg/L. Eccles et al. [49] and Schuijt et al. [48] report two cut-off values, 20 mg/L and 100 mg/L. Thus, the range that the ImmunoDisk covers is compatible with these application-specific and clinically relevant cut-off values. Additionally, these concentration ranges and LOD values obtained using the fully integrated ImmunoDisk are in good agreement with the obtained range and LOD when the BFPD-IA was conducted on a benchtop laboratory instrument [46]. Furthermore, the TAT of the ImmunoDisk for the detection of CRP is 20 min and three samples can be processed in parallel.

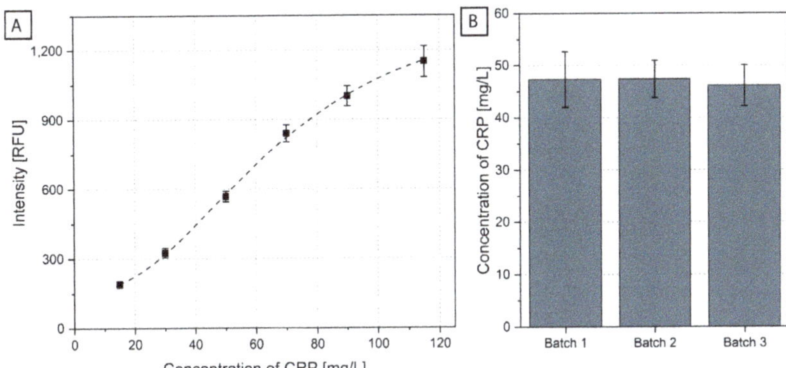

Figure 4. (**A**) A calibration curve (N = 6 for 50 mg/L, N = 7 for all other concentrations) for the system was obtained by running several concentrations on disks produced in three different batches. It is normalized to the overall blank (results of each batch are shown in Supplementary File S10). (**B**) Averages and standard deviations of the CRP concentration in the CRM sample were measured using three batches of disks (N = 11 measurements in total, batch 1 N = 4, batch 2 N = 3, batch 3 N = 4). The calculation of the CRP concentration was based on the calibration curve of (**A**) and the 4-parameter fit shown in Supplementary File S10. The 11 CRM measurements are shown in Supplementary File S11.

The obtained calibration curve (information on the fit of the curve in Supplementary File S10) was used to calculate the concentration of CRP in a CRM (European Commission Joint Research Centre, Geel, Belgium; ERM-DA474/IFCC). Three different batches of disks were again produced and were used for the measurement of CRM (Figure 4B). The CRM was measured 12 times. One measurement led to an outlier (see Supplementary File S11 for the calculation of the outlier), which could stem from handling variations in one or more steps during the currently manual production of the disks, including the insertion of the stickpack, the pipetting of the beads or the sealing of the disk. These sources of errors will be reduced when the cartridge is scaled up to mass production (notably, all steps of the manufacturing process are compatible with an automated production line [74]). The expected value of the CRM was 41.2 $\pm$ 2.5 mg/L, measured using immunonephelometry and immunoturbidimetry, without pre-stored reagents [75]. The average measured value on the ImmunoDisk was 46.9 $\pm$ 4.0 mg/L (N = 11). A deviation between measurements with different systems and detection principles can be expected. For diagnostic applications where CRP is used as biomarker to support antibiotic stewardship, it is sufficient to provide the CRP concentration between/beyond cut-off values that are reported in clinical studies and guidelines [48,49,72,73]. In any case, the usability of the test for diagnostic purposes needs to be evaluated in a clinical study, which is outside the scope of this work. In the context of our technical feasibility assessment, the inter-disk concentration CV in our case was only 8.6% (calculated over all batches). Moreover, the three different ImmunoDisk batches showed great reproducibility, with an inter-batch CV of 1.5% calculated using the average measured concentration of each batch.

3.9. Overall Assessment of the ImmunoDisk

The overall goal was to fully integrate (including pre-storage of reagents) the BFPD-IA on a microfluidic cartridge. Notably, the initial version of the BFPD-IA, conducted in a microtiter plate, included a step to transfer the bound-free phase to a separate well for detection after the separation step. However, through the interplay between microfluidic design and protocols, we managed to not only reduce the manual handling steps to just one (insertion of the sample), but also to omit the transfer of the bound-free phase and to achieve incubation, separation and detection all in the multipurpose chamber. Thus, the few and easy steps of the BFPD-IA itself [46], together with smart microfluidic solutions, enabled a simple transfer of the assay to a space-saving microfluidic design. This also has a potential impact on the test throughput. Specifically, in our case, three identical structures (including the blood–plasma separation module that was not used in this work) could fit on a single disk, which means that in the future, three patients can be screened simultaneously. The simple structure of the disk, which utilizes a multipurpose chamber approach for multiple operations, means that no additional sensors, extra components or surface treatment are required for integration of the BFPD-IA method. This demonstrates how the simplicity of a biochemical assay protocol like the BFPD-IA, which has a single step, is wash-free and rapid, can be crucial for the sustainability, viability, robustness and cost of a PoC system.

For the full integration of a bead-based immunoassay PoC system like the one we propose, it is inevitable that the pre-storage of protein-coupled beads comes into focus. The parametric study of this topic presented in this publication revealed some interesting findings.

We show for the first time, to the best of our knowledge, the combination of a polymeric and a disaccharide as additives for the pre-storage of protein-coupled beads using air drying. We observed that the buffer with a combination of PEG1000 and trehalose performed best for the CRP antibody-coated magnetic beads, but that PEG itself proved to be incompatible with the CRP antigen-coated fluorescent beads, most probably due to the different nature of the coupled protein. The common drying protocol for the two different types of beads and proteins on the same disk is also an important outcome. This can open the way for other bead-based assays to be integrated in microfluidic systems.

In its fully integrated version, the ImmunoDisk is able to measure high concentrations of CRP in a 5 μL human serum sample, which is inserted directly into the cartridge with no prior dilution steps and no washing steps on disk. These conditions simplify the fluidic integration and reduce the footprint on disk. Importantly, the ImmunoDisk was processed on a device, the LabDisk Player 1 functional model, which was developed for the automation of nucleic acid amplification technologies (NAATs, e.g., PCR and isothermal methods) and for which different applications have already been shown (e.g., respiratory tract infections [64], tropical infections [63], oral diseases [62] and vector analysis in mosquitos [76]), without any changes to the device hardware. The compatibility of the ImmunoDisk with this NAAT device is primarily due to the protocol and the detection principle of the BFPD-IA, as well as the adaptation of the microfluidic structures. This may prospectively pave the way for the more general implementation of the BFPD-IA on other NAAT devices, thereby expanding their portfolio and achieving complementary diagnostics where, for example, pathogen identification and immunological response are tested on the same instrument, with major impacts on the health economics of diagnostics. Moreover, the health systems in several areas of medicine that require such co-assessment and interoperability between microbiological and immunological outcomes [50,73,77–80] would benefit from such complementary diagnostics.

4. Conclusions and Outlook

In this work, we show, for the first time, the full sample-to-answer integration and automation of a heterogeneous wash-free, bead-based, bound-free phase detection immunoassay for the detection of CRP in serum in the clinically relevant range, including in situ detection and pre-storage of all involved reagents. The results show the great potential of the ImmunoDisk to be processed on the same centrifugal microfluidic processing device that performs PCR or isothermal amplification.

For the pre-storage of the protein-coupled beads, we investigated all the key parameters for the air drying of proteins on particles and gained some key insights on the influence of additives and the importance of the drying temperature and duration combination. Thus, this work contributes significantly to the research methodology around the pre-storage of beads coupled with proteins, for which there is, to the best of our knowledge, limited literature available, unlike for proteins or beads alone. From the microfluidic perspective on the pre-storage of protein-coupled beads, we propose using a storage chamber on the cartridge for a more reproducible and robust storage and release of the protein-coupled beads. Furthermore, a multipurpose chamber is introduced that facilitates the assay incubation, separation and detection, leading to a simplified, robust, space-saving microfluidic design. The future outlook for this work includes the expansion of the pre-storage protocols to other protein assays, covering applications such as oral health, cardiovascular diseases or sepsis, and the implementation of the CRP-ImmunoDisk with clinical samples and potentially in combination with molecular diagnostics for improved patient management at the PoC.

Supplementary Materials: The following supporting information can be downloaded at: https://www.mdpi.com/article/10.3390/bios12060413/s1, File S1: ImmunoDisk—fluidic protocol; File S2: Sedimentation of magnetic beads; File S3: Calculation of the distance traveled by a bead at specific frequencies; File S4: Results of parametric study for the drying of magnetic beads; File S5: Magnetic bead resuspension experiments on disk; File S6: Overview of different drying buffers for native CRP-coupled fluorescent beads; File S7: MMP-9 assay with dried antibody-coated magnetic beads and liquid antigen-coated fluorescent beads; File S8: Design of Experiments (DoE); File S9: Overview of ΔIntensity and CV using different drying temperatures and durations; File S10: Calibration curve; File S11: Data of CRM measurements and outlier calculation.

Author Contributions: Conceptualization, B.J., D.B. and L.K.; methodology, B.J.; validation, B.J.; formal analysis, B.J.; investigation, B.J. and L.K.; resources, M.K.; writing—original draft preparation, B.J.; writing—review and editing, D.B., L.K., N.P., M.K., N.B., R.Z. and K.M.; visualization, B.J.; supervision, B.J., D.B., N.B., R.Z. and K.M.; project administration, B.J. and K.M.; funding acquisition, N.B. and K.M. All authors have read and agreed to the published version of the manuscript.

Funding: This research was funded by the European Union's Horizon 2020 research and innovation program under grant agreement no. 633780 ('DIAGORAS' project) and by the project 'Respiotic' (E!113595), which is carried out within the framework of the European funding program 'Eurostars'. The German partner of 'Respiotic' was funded by the Federal Ministry of Education and Research (FZK: 01QE1939B). We also acknowledge support by the Open Access Publication Fund of the University of Freiburg.

Institutional Review Board Statement: Not applicable.

Informed Consent Statement: Not applicable.

Data Availability Statement: Data are contained within the article or in the Supplementary Material.

Acknowledgments: The authors would like to acknowledge the Hahn-Schickard Lab-on-a-Chip Foundry Service and especially Philip Koch and Philipp Tepper for LabDisk and stickpack production. We also thank Lara Müller for her insights.

Conflicts of Interest: The authors declare the following competing financial interests: B.J. and K.M. have filed a patent on the BFPD immunoassay that is used in this application. The funders had no role in the design of the study; in the collection, analyses, or interpretation of data; in the writing of the manuscript, or in the decision to publish the results.

References

1. Luong, J.; Vashist, S.K. (Eds.) *Handbook of Immunoassay Technologies: Approaches, Performances, and Applications*; Academic Press an imprint of Elsevier: London, UK, 2018.
2. Gug, I.T.; Tertis, M.; Hosu, O.; Cristea, C. Salivary biomarkers detection: Analytical and immunological methods overview. *TrAC-Trends Anal. Chem.* **2019**, *113*, 301–316. [CrossRef]
3. Silbereisen, A.; Alassiri, S.; Bao, K.; Grossmann, J.; Nanni, P.; Fernandez, C.; Tervahartiala, T.; Nascimento, G.G.; Belibasakis, G.N.; Heikkinen, A.-M.; et al. Label-Free Quantitative Proteomics versus Antibody-Based Assays to Measure Neutrophil-Derived Enzymes in Saliva. *Proteom. Clin. Appl.* **2020**, *14*, e1900050. [CrossRef] [PubMed]
4. Hnasko, R. *ELISA: Methods and Protocols*; Humana Press: New York, NY, USA; Heidelberg, Germany; Dordrecht, The Netherlands; London, UK, 2015.
5. O'Kennedy, R.; Murphy, C. *Immunoassays: Development, Applications and Future Trends*; Pan Stanford Publishing: Milton, GA, USA, 2017.
6. Bostanci, N.; Mitsakakis, K.; Afacan, B.; Bao, K.; Johannsen, B.; Baumgartner, D.; Müller, L.; Kotolová, H.; Emingil, G.; Karpíšek, M. Validation and verification of predictive salivary biomarkers for oral health. *Sci. Rep.* **2021**, *11*, 6406. [CrossRef] [PubMed]

7. Lhopitallier, L.; Kronenberg, A.; Meuwly, J.-Y.; Locatelli, I.; Mueller, Y.; Senn, N.; D'Acremont, V.; Boillat-Blanco, N. Procalcitonin and lung ultrasonography point-of-care testing to determine antibiotic prescription in patients with lower respiratory tract infection in primary care: Pragmatic cluster randomised trial. *BMJ* **2021**, *374*, n2132. [CrossRef]
8. Teggert, A.; Datta, H.; Ali, Z. Biomarkers for Point-of-Care Diagnosis of Sepsis. *Micromachines* **2020**, *11*, 286. [CrossRef]
9. Bostanci, N.; Selevsek, N.; Wolski, W.; Grossmann, J.; Bao, K.; Wahlander, A.; Trachsel, C.; Schlapbach, R.; Öztürk, V.Ö.; Afacan, B.; et al. Targeted Proteomics Guided by Label-free Quantitative Proteome Analysis in Saliva Reveal Transition Signatures from Health to Periodontal Disease. *Mol. Cell. Proteom.* **2018**, *17*, 1392–1409. [CrossRef]
10. Lim, C.T.; Zhang, Y. Bead-based microfluidic immunoassays: The next generation. *Biosens. Bioelectron.* **2007**, *22*, 1197–1204. [CrossRef]
11. Pecoraro, V.; Banfi, G.; Germagnoli, L.; Trenti, T. A systematic evaluation of immunoassay point-of-care testing to define impact on patients' outcomes. *Ann. Clin. Biochem.* **2017**, *54*, 420–431. [CrossRef]
12. de Oliveira, V.M.; Moraes, R.B.; Stein, A.T.; Wendland, E.M. Accuracy of C-Reactive protein as a bacterial infection marker in critically immunosuppressed patients: A systematic review and meta-analysis. *J. Crit. Care* **2017**, *42*, 129–137. [CrossRef]
13. Strohmeier, O.; Keller, M.; Schwemmer, F.; Zehnle, S.; Mark, D.; von Stetten, F.; Zengerle, R.; Paust, N. Centrifugal microfluidic platforms: Advanced unit operations and applications. *Chem. Soc. Rev.* **2015**, *44*, 6187–6229. [CrossRef]
14. Maguire, I.; O'Kennedy, R.; Ducrée, J.; Regan, F. A review of centrifugal microfluidics in environmental monitoring. *Anal. Methods* **2018**, *10*, 1497–1515. [CrossRef]
15. Zehnle, S.; Rombach, M.; Zengerle, R.; von Stetten, F.; Paust, N. Network simulation-based optimization of centrifugo-pneumatic blood plasma separation. *Biomicrofluidics* **2017**, *11*, 24114. [CrossRef] [PubMed]
16. Johannsen, B.; Müller, L.; Baumgartner, D.; Karkossa, L.; Früh, S.M.; Bostanci, N.; Karpíšek, M.; Zengerle, R.; Paust, N.; Mitsakakis, K. Automated Pre-Analytic Processing of Whole Saliva Using Magnet-Beating for Point-of-Care Protein Biomarker Analysis. *Micromachines* **2019**, *10*, 833. [CrossRef] [PubMed]
17. Czilwik, G.; Vashist, S.K.; Klein, V.; Buderer, A.; Roth, G.; von Stetten, F.; Zengerle, R.; Mark, D. Magnetic chemiluminescent immunoassay for human C-reactive protein on the centrifugal microfluidics platform. *RSC Adv.* **2015**, *5*, 61906–61912. [CrossRef]
18. Hemmi, A.; Usui, T.; Moto, A.; Tobita, T.; Soh, N.; Nakano, K.; Zeng, H.; Uchiyama, K.; Imato, T.; Nakajima, H. A surface plasmon resonance sensor on a compact disk-type microfluidic device. *J. Sep. Sci.* **2011**, *34*, 2913–2919. [CrossRef]
19. Honda, N.; Lindberg, U.; Andersson, P.; Hoffmann, S.; Takei, H. Simultaneous multiple immunoassays in a compact disc-shaped microfluidic device based on centrifugal force. *Clin. Chem.* **2005**, *51*, 1955–1961. [CrossRef]
20. Hosseini, S.; Aeinehvand, M.M.; Uddin, S.M.; Benzina, A.; Rothan, H.A.; Yusof, R.; Koole, L.H.; Madou, M.J.; Djordjevic, I.; Ibrahim, F. Microsphere integrated microfluidic disk: Synergy of two techniques for rapid and ultrasensitive dengue detection. *Sci. Rep.* **2015**, *5*, 16485. [CrossRef]
21. Kim, T.-H.; Abi-Samra, K.; Sunkara, V.; Park, D.-K.; Amasia, M.; Kim, N.; Kim, J.; Kim, H.; Madou, M.; Cho, Y.-K. Flow-enhanced electrochemical immunosensors on centrifugal microfluidic platforms. *Lab Chip* **2013**, *13*, 3747–3754. [CrossRef]
22. Lai, S.; Wang, S.; Luo, J.; Lee, L.J.; Yang, S.-T.; Madou, M.J. Design of a compact disk-like microfluidic platform for enzyme-linked immunosorbent assay. *Anal. Chem.* **2004**, *76*, 1832–1837. [CrossRef]
23. Lee, B.S.; Lee, J.-N.; Park, J.-M.; Lee, J.-G.; Kim, S.; Cho, Y.-K.; Ko, C. A fully automated immunoassay from whole blood on a disc. *Lab Chip* **2009**, *9*, 1548–1555. [CrossRef]
24. Lin, Q.; Wu, J.; Fang, X.; Kong, J. Washing-free centrifugal microchip fluorescence immunoassay for rapid and point-of-care detection of protein. *Anal. Chim. Acta* **2020**, *1118*, 18–25. [CrossRef] [PubMed]
25. Lutz, S.; Lopez-Calle, E.; Espindola, P.; Boehm, C.; Brueckner, T.; Spinke, J.; Marcinowski, M.; Keller, T.; Tgetgel, A.; Herbert, N.; et al. A fully integrated microfluidic platform for highly sensitive analysis of immunochemical parameters. *Analyst* **2017**, *142*, 4206–4214. [CrossRef] [PubMed]
26. Nwankire, C.E.; Donohoe, G.G.; Zhang, X.; Siegrist, J.; Somers, M.; Kurzbuch, D.; Monaghan, R.; Kitsara, M.; Burger, R.; Hearty, S.; et al. At-line bioprocess monitoring by immunoassay with rotationally controlled serial siphoning and integrated supercritical angle fluorescence optics. *Anal. Chim. Acta* **2013**, *781*, 54–62. [CrossRef] [PubMed]
27. Park, J.; Sunkara, V.; Kim, T.-H.; Hwang, H.; Cho, Y.-K. Lab-on-a-disc for fully integrated multiplex immunoassays. *Anal. Chem.* **2012**, *84*, 2133–2140. [CrossRef]
28. Park, Y.-S.; Sunkara, V.; Kim, Y.; Lee, W.S.; Han, J.-R.; Cho, Y.-K. Fully Automated Centrifugal Microfluidic Device for Ultrasensitive Protein Detection from Whole Blood. *J. Vis. Exp.* **2016**, *110*, 54143. [CrossRef]
29. Riegger, L.; Grumann, M.; Nann, T.; Riegler, J.; Ehlert, O.; Bessler, W.; Mittenbuehler, K.; Urban, G.; Pastewka, L.; Brenner, T.; et al. Read-out concepts for multiplexed bead-based fluorescence immunoassays on centrifugal microfluidic platforms. *Sens. Actuators A-Phys.* **2006**, *126*, 455–462. [CrossRef]
30. Schaff, U.Y.; Sommer, G.J. Whole blood immunoassay based on centrifugal bead sedimentation. *Clin. Chem.* **2011**, *57*, 753–761. [CrossRef]
31. Shih, C.-H.; Wu, H.-C.; Chang, C.-Y.; Huang, W.-H.; Yang, Y.-F. An enzyme-linked immunosorbent assay on a centrifugal platform using magnetic beads. *Biomicrofluidics* **2014**, *8*, 52110. [CrossRef]
32. Uddin, R.; Donolato, M.; Hwu, E.-T.; Hansen, M.F.; Boisen, A. Combined detection of C-reactive protein and PBMC quantification from whole blood in an integrated lab-on-a-disc microfluidic platform. *Sens. Actuators B-Chem.* **2018**, *272*, 634–642. [CrossRef]

33. Wang, K.; Liang, R.; Chen, H.; Lu, S.; Jia, S.; Wang, W. A microfluidic immunoassay system on a centrifugal platform. *Sens. Actuators B-Chem.* **2017**, *251*, 242–249. [CrossRef]
34. Wu, H.-C.; Chen, Y.-H.; Shih, C.-H. Disk-based enzyme-linked immunosorbent assays using the liquid-aliquoting and siphoning-evacuation technique. *Biomicrofluidics* **2018**, *12*, 54101. [CrossRef] [PubMed]
35. Zhao, Y.; Czilwik, G.; Klein, V.; Mitsakakis, K.; Zengerle, R.; Paust, N. C-reactive protein and interleukin 6 microfluidic immunoassays with on-chip pre-stored reagents and centrifugo-pneumatic liquid control. *Lab Chip* **2017**, *17*, 1666–1677. [CrossRef]
36. Abe, T.; Okamoto, S.; Taniguchi, A.; Fukui, M.; Yamaguchi, A.; Utsumi, Y.; Ukita, Y. A lab in a bento box: An autonomous centrifugal microfluidic system for an enzyme-linked immunosorbent assay. *Anal. Methods* **2020**, *12*, 4858–4866. [CrossRef] [PubMed]
37. Lee, W.; Jung, J.; Hahn, Y.K.; Kim, S.K.; Lee, Y.; Lee, J.; Lee, T.-H.; Park, J.-Y.; Seo, H.; Lee, J.N.; et al. A centrifugally actuated point-of-care testing system for the surface acoustic wave immunosensing of cardiac troponin I. *Analyst* **2013**, *138*, 2558–2566. [CrossRef] [PubMed]
38. Burger, R.; Reith, P.; Kijanka, G.; Akujobi, V.; Abgrall, P.; Ducrée, J. Array-based capture, distribution, counting and multiplexed assaying of beads on a centrifugal microfluidic platform. *Lab Chip* **2012**, *12*, 1289–1295. [CrossRef]
39. Miyazaki, C.M.; Kinahan, D.J.; Mishra, R.; Mangwanya, F.; Kilcawley, N.; Ferreira, M.; Ducrée, J. Label-free, spatially multiplexed SPR detection of immunoassays on a highly integrated centrifugal Lab-on-a-Disc platform. *Biosens. Bioelectron.* **2018**, *119*, 86–93. [CrossRef]
40. Gao, Z.; Chen, Z.; Deng, J.; Li, X.; Qu, Y.; Xu, L.; Luo, Y.; Lu, Y.; Liu, T.; Zhao, W.; et al. Measurement of Carcinoembryonic Antigen in Clinical Serum Samples Using a Centrifugal Microfluidic Device. *Micromachines* **2018**, *9*, 470. [CrossRef]
41. Gijs, M.A.M.; Lacharme, F.; Lehmann, U. Microfluidic applications of magnetic particles for biological analysis and catalysis. *Chem. Rev.* **2010**, *110*, 1518–1563. [CrossRef]
42. Tighe, P.J.; Ryder, R.R.; Todd, I.; Fairclough, L.C. ELISA in the multiplex era: Potentials and pitfalls. *Proteom. Clin. Appl.* **2015**, *9*, 406–422. [CrossRef]
43. Diamandis, E.P.; Christopoulos, T.K. *Immunoassay*; Academic Press: San Diego, CA, USA, 1996.
44. Slagle, K.M.; Ghosn, S.J. Immunoassays: Tools for Sensitive, Specific, and Accurate Test Results. *Lab. Med.* **1996**, *27*, 177–183. [CrossRef]
45. Mensink, M.A.; Frijlink, H.W.; van der Voort Maarschalk, K.; Hinrichs, W.L.J. How sugars protect proteins in the solid state and during drying (review): Mechanisms of stabilization in relation to stress conditions. *Eur. J. Pharm. Biopharm.* **2017**, *114*, 288–295. [CrossRef] [PubMed]
46. Johannsen, B.; Karpíšek, M.; Baumgartner, D.; Klein, V.; Bostanci, N.; Paust, N.; Früh, S.M.; Zengerle, R.; Mitsakakis, K. One-step, wash-free, bead-based immunoassay employing bound-free phase detection. *Anal. Chim. Acta* **2021**, *1153*, 338280. [CrossRef] [PubMed]
47. Marnell, L.; Mold, C.; Du Clos, T.W. C-reactive protein: Ligands, receptors and role in inflammation. *Clin. Immunol.* **2005**, *117*, 104–111. [CrossRef] [PubMed]
48. Schuijt, T.J.; Boss, D.S.; Musson, R.E.A.; Demir, A.Y. Influence of point-of-care C-reactive protein testing on antibiotic prescription habits in primary care in the Netherlands. *Fam. Pract.* **2018**, *35*, 179–185. [CrossRef]
49. Eccles, S.; Pincus, C.; Higgins, B.; Woodhead, M. Diagnosis and management of community and hospital acquired pneumonia in adults: Summary of NICE guidance. *BMJ* **2014**, *349*, g6722. [CrossRef]
50. Johannsen, B.; Mark, D.; Boillat-Blanco, N.; Fresco, A.; Baumgartner, D.; Zengerle, R.; Mitsakakis, K. Rapid Diagnosis of Respiratory Tract Infections Using a Point-of-Care Platform Incorporating a Clinical Decision Support Algorithm. *Stud. Health Technol. Inform.* **2020**, *273*, 234–239.
51. Prins, H.J.; Duijkers, R.; van der Valk, P.; Schoorl, M.; Daniels, J.M.A.; van der Werf, T.S.; Boersma, W.G. CRP-guided antibiotic treatment in acute exacerbations of COPD in hospital admissions. *Eur. Respir. J.* **2019**, *53*, 1802014. [CrossRef]
52. Schwarz, I.; Zehnle, S.; Hutzenlaub, T.; Zengerle, R.; Paust, N. System-level network simulation for robust centrifugal-microfluidic lab-on-a-chip systems. *Lab Chip* **2016**, *16*, 1873–1885. [CrossRef]
53. Focke, M.; Stumpf, F.; Faltin, B.; Reith, P.; Bamarni, D.; Wadle, S.; Müller, C.; Reinecke, H.; Schrenzel, J.; Francois, P.; et al. Microstructuring of polymer films for sensitive genotyping by real-time PCR on a centrifugal microfluidic platform. *Lab Chip* **2010**, *10*, 2519–2526. [CrossRef]
54. van Oordt, T.; Barb, Y.; Smetana, J.; Zengerle, R.; von Stetten, F. Miniature stick-packaging—an industrial technology for pre-storage and release of reagents in lab-on-a-chip systems. *Lab Chip* **2013**, *13*, 2888–2892. [CrossRef]
55. Zehnle, S.; Schwemmer, F.; Roth, G.; von Stetten, F.; Zengerle, R.; Paust, P. Centrifugo-dynamic inward pumping of liquids on a centrifugal microfluidic platform. *Lab Chip* **2012**, *12*, 5142–5145. [CrossRef] [PubMed]
56. Hess, J.F.; Zehnle, S.; Juelg, P.; Hutzenlaub, T.; Zengerle, R.; Paust, N. Review on pneumatic operations in centrifugal microfluidics. *Lab Chip* **2019**, *19*, 3745–3770. [CrossRef] [PubMed]
57. Grumann, M.; Geipel, A.; Riegger, L.; Zengerle, R.; Ducrée, J. Batch-mode mixing on centrifugal microfluidic platforms. *Lab Chip* **2005**, *5*, 560–565. [CrossRef] [PubMed]
58. Li, Q.; Luo, K.H.; Kang, Q.J.; Chen, Q. Contact angles in the pseudopotential lattice Boltzmann modeling of wetting. *Phys. Rev. E* **2014**, *90*, 53301. [CrossRef] [PubMed]

59. Jacobsen, A.E.; Sullivan, W.F. Centrifugal Sedimentation Method for Particle Size Distribution. *Ind. Eng. Chem. Anal. Ed.* **1946**, *18*, 360–364. [CrossRef]
60. Scott, D.J.; Harding, S.E.; Rowe, A.J.; Scott, D.; Rowe, A. (Eds.) Introduction to Differential Sedimentation. In *Analytical Ultracentrifugation*; Royal Society of Chemistry: Cambridge, UK, 2007; pp. 270–290.
61. Manning, M.C.; Chou, D.K.; Murphy, B.M.; Payne, R.W.; Katayama, D.S. Stability of protein pharmaceuticals: An update. *Pharm. Res.* **2010**, *27*, 544–575. [CrossRef]
62. Baumgartner, D.; Johannsen, B.; Specht, M.; Lüddecke, J.; Rombach, M.; Hin, S.; Paust, N.; von Stetten, F.; Zengerle, R.; Herz, C.; et al. OralDisk: A Chair-Side Compatible Molecular Platform Using Whole Saliva for Monitoring Oral Health at the Dental Practice. *Biosensors* **2021**, *11*, 423. [CrossRef]
63. Hin, S.; Lopez-Jimena, B.; Bakheit, M.; Klein, V.; Stack, S.; Fall, C.; Sall, A.; Enan, K.; Mustafa, M.; Gillies, L.; et al. Fully automated point-of-care differential diagnosis of acute febrile illness. *PLoS Negl. Trop. Dis.* **2021**, *15*, e0009177. [CrossRef]
64. Rombach, M.; Hin, S.; Specht, M.; Johannsen, B.; Lüddecke, J.; Paust, N.; Zengerle, R.; Roux, L.; Sutcliffe, T.; Peham, J.R.; et al. RespiDisk: A point-of-care platform for fully automated detection of respiratory tract infection pathogens in clinical samples. *Analyst* **2020**, *145*, 7040–7047. [CrossRef]
65. Molecular Probes, FluoSpheres®Fluorescent Microspheres: Product Information. 2005. Available online: https://assets.thermofisher.com/TFS-Assets/LSG/manuals/mp05000.pdf (accessed on 11 May 2022).
66. Jeyachandran, Y.L.; Mielczarski, J.A.; Mielczarski, E.; Rai, B. Efficiency of blocking of non-specific interaction of different proteins by BSA adsorbed on hydrophobic and hydrophilic surfaces. *J. Colloid Interface Sci.* **2010**, *341*, 136–142. [CrossRef]
67. Lee, J.C.; Lee, L.L. Preferential solvent interactions between proteins and polyethylene glycols. *J. Biol. Chem.* **1981**, *256*, 625–631. [CrossRef]
68. Arakawa, T.; Timasheff, S.N. Mechanism of poly(ethylene glycol) interaction with proteins. *Biochemistry* **1985**, *24*, 6756–6762. [CrossRef] [PubMed]
69. Bhat, R.; Timasheff, S.N. Steric exclusion is the principal source of the preferential hydration of proteins in the presence of polyethylene glycols. *Protein Sci.* **1992**, *1*, 1133–1143. [CrossRef] [PubMed]
70. Rawat, S.; Raman Suri, C.; Sahoo, D.K. Molecular mechanism of polyethylene glycol mediated stabilization of protein. *Biochem. Biophys. Res. Commun.* **2010**, *392*, 561–566. [CrossRef] [PubMed]
71. Chakraborty, C.; Agrawal, A. Computational analysis of C-reactive protein for assessment of molecular dynamics and interaction properties. *Cell Biochem. Biophys.* **2013**, *67*, 645–656. [CrossRef]
72. Jakobsen, K.A.; Melbye, H.; Kelly, M.J.; Ceynowa, C.; Mölstad, S.; Hood, K.; Butler, C.C. Influence of CRP testing and clinical findings on antibiotic prescribing in adults presenting with acute cough in primary care. *Scand. J. Prim. Health Care* **2010**, *28*, 229–236. [CrossRef]
73. Alcoba, G.; Keitel, K.; Maspoli, V.; Lacroix, L.; Manzano, S.; Gehri, M.; Tabin, R.; Gervaix, A.; Galetto-Lacour, A. A three-step diagnosis of pediatric pneumonia at the emergency department using clinical predictors, C-reactive protein, and pneumococcal PCR. *Eur. J. Pediatr.* **2017**, *176*, 815–824. [CrossRef]
74. Hahn-Schickard-Gesellschaft für Angewandte Forschung e.V., Lab-on-a-Chip Foundry. Available online: https://www.hahn-schickard.de/en/service-portfolio/production/lab-on-a-chip-foundry (accessed on 11 May 2022).
75. Auclair, G.; Zegers, I.; Charoud-Got, J.; Munoz-Pineiro, M.; Hanisch, K.; Boulo, S.; Trapmann, S.; Schimmel, H.; Emons, H.; Schreiber, W. *The Certification of the Mass Concentration of C-Reactive Protein in Human Serum—Certified Reference Material ERM®-DA474/IFCC*; EUR 24922 EN; Publications Office of the European Union: Luxembourg, 2011.
76. Hin, S.; Baumgartner, D.; Specht, M.; Lüddecke, J.; Mahmodi Arjmand, E.; Johannsen, B.; Schiedel, L.; Rombach, M.; Paust, N.; von Stetten, F.; et al. VectorDisk: A Microfluidic Platform Integrating Diagnostic Markers for Evidence-Based Mosquito Control. *Processes* **2020**, *8*, 1677. [CrossRef]
77. Mitsakakis, K. Novel lab-on-a-disk platforms: A powerful tool for molecular fingerprinting of oral and respiratory tract infections. *Expert Rev. Mol. Diagn.* **2021**, *21*, 523–526. [CrossRef]
78. Teh, R.; Tee, W.D.; Tan, E.; Fan, K.; Koh, C.J.; Tambyah, P.A.; Oon, J.; Tee, N.; Soh, A.Y.S.; Siah, K.T.H. Review of the role of gastrointestinal multiplex polymerase chain reaction in the management of diarrheal illness. *J. Gastroenterol. Hepatol.* **2021**, *36*, 3286–3297. [CrossRef]
79. Chen, W.; Zheng, K.I.; Liu, S.; Yan, Z.; Xu, C.; Qiao, Z. Plasma CRP level is positively associated with the severity of COVID-19. *Ann. Clin. Microbiol. Antimicrob.* **2020**, *19*, 18. [CrossRef] [PubMed]
80. Chen, C.-C.; Lee, I.-K.; Liu, J.-W.; Huang, S.-Y.; Wang, L. Utility of C-Reactive Protein Levels for Early Prediction of Dengue Severity in Adults. *Biomed. Res. Int.* **2015**, *2015*, 936062. [CrossRef] [PubMed]

 biosensors MDPI

Article

Rapid Capturing and Chemiluminescent Sensing of Programmed Death Ligand-1 Expressing Extracellular Vesicles

Adeel Khan [1], Kaili Di [2], Haroon Khan [3], Nongyue He [1,*] and Zhiyang Li [2]

1 State Key Laboratory of Bioelectronics, School of Biological Science and Medical Engineering, National Demonstration Center for Experimental Biomedical Engineering Education, Southeast University, Nanjing 210096, China; adeel9925@seu.edu.cn
2 Department of Clinical Laboratory Medicine, the Affiliated Drum Tower Hospital, Medical School of Nanjing University, Nanjing 210008, China; mg1835040@smail.nju.edu.cn (K.D.); lizhiyang@nju.edu.cn (Z.L.)
3 Neuroscience and Neuroengineering Research Center, Med-X Research Institute, School of Biomedical Engineering, Shanghai Jiao Tong University, Shanghai 200240, China; harry_884489@sjtu.edu.cn
* Correspondence: nyhe@seu.edu.cn

Abstract: Cancer specific extracellular vesicles (EVs) are of significant clinical relevance, for instance programmed death ligand-1 (PD-L1) expressing EVs (PD-L1@EVs) have been shown to be ideal biomarker for non-invasive diagnosis of cancer and can predate the response of cancer patients to anti-PD-1/PD-L-1 immunotherapy. The development of sensitive and straightforward methods for detecting PD-L1@EVs can be a vital tool for non-invasive diagnosis of cancer. Most of the contemporary methods for EVs detection have limitations such as involvement of long and EV's loss prone isolation methods prior to detection or they have employed expensive antibodies and instruments to accomplish detection. Therefore, we designed an ultracentrifugation-free and antibody-free sensing assay for PD-L1@EV by integrating Titanium oxide (TiO_2) coated magnetic beads (Fe_3O_4@TiO_2) rapid capturing of EVs from undiluted serum with aptamers specificity and chemiluminescence (CL) sensitivity. To accomplish this we used Fe_3O_4@TiO_2 beads to rapidly capture EVs from the undiluted patient serum and added biotin labelled PD-L1 aptamer to specifically recognize PD-L1@EVs. Later, added streptavidin-modified Alkaline phosphates (ALP) taking advantage of biotin-streptavidin strong binding. Addition of CDP-star, a chemiluminescent substrate of ALP, initiates the chemiluminiscense that was recorded using spectrophotometer. The sensing assay showed high sensitivity with limit of detection (LOD) as low as 2.584×10^5 EVs/mL and a wider linear correlation of CL intensity (a.u.) with the concentration of PD-L1@EVs from 10^5 to 10^8 EVs/mL. To examine the clinical utility of sensing assay we used undiluted serum samples from lung cancer patients and healthy individuals and successfully discern between healthy individuals and lung cancer patients. We are optimistic that the sensing assay can ameliorate our ability to be able to diagnose lung cancer non-invasively and can be helpful to predate the patient's response to anti-PD-1/PD-L1 immunotherapy.

Keywords: extracellular vesicles; Fe_3O_4@TiO_2; PD-L1; aptasensor; chemiluminescence; lung cancer; non-invasive diagnosis

Citation: Khan, A.; Di, K.; Khan, H.; He, N.; Li, Z. Rapid Capturing and Chemiluminescent Sensing of Programmed Death Ligand-1 Expressing Extracellular Vesicles. *Biosensors* **2022**, *12*, 281. https://doi.org/10.3390/bios12050281

Received: 16 March 2022
Accepted: 26 April 2022
Published: 28 April 2022

1. Introduction

Extracellular vesicles (EVs) are membrane enclosed nano-size-entities (50–1000 nm diameter) with a biologically active cargo. A plethora of cell types release EVs. The presence of EVs has been reported in majority of body fluids for instance blood, urine, saliva and milk, hence, can be amassed non-invasively [1]. Cancer cells release significantly more EVs than normal cells. A growing body of evidence propose that EVs released by cancer cells can be diagnostic markers for a various cancers, valuable for cancer monitoring, early detection of cancer relapse, and even response to an anti-cancer therapy [2]. One such example are the EVs expressing Programmed death ligand-1 (PD-L1). PD-L1 also known as CD274 is a crucial immune check point protein that has been reported to be expressed on the surface of cancer cells as well as cancer cells derived EVs [3]. EVs with PD-L1 marker (PD-L1@EVs) can cause strong immunosuppressive effects [4]. To accomplish immunosuppression the PD-L1@EVs binds the activated T cells via PD-1 receptor resulting in cessation of their proliferation, cytokine production and cytotoxicity. PD-L1@EVs can be diagnostic markers and can also stratify between patients responding and non-responding to anti-PD-1/PD-L1 immunotherapy [3,5].

Lung cancer (LC) is one of the deadliest cancers both in males and females with a dismal 5-years survival rate. The new cases of LC are expanding with alarming rate [6]. Mostly LC patients are diagnosed at late stages, when the treatment options are narrow. It is therefore crucial to develop techniques that aid in the non-invasive early diagnosis of LC; a stage more amenable to treatment. EVs' abundance in body fluids and non-invasive collection confer EVs a status of ideal biomarker for diagnosis of LC [7]. To develop excellent detection method for EVs an important step is efficient and rapid capturing of EVs from clinical samples. EVs amassment by ultracentrifugation/commercial kits is a laborious and long operation (from several hours to overnight). Ultrafiltration based isolation of EVs is simple and fast but high loss of EVs limits its utility. Polymer based EVs precipitation is quite simple but low specificity, and co-enrichment of contaminates with EVs discourage its use [8]. These canonical limitations of traditional methods intensify the need for better enrichment strategies [9]. Advance methods such as acoustofluidic methods yield EVs comparable to ultracentrifugation by using a very small sample volume, however research is still needed to optimize the combined acoustic/microfluidic forces for efficient control of nanoparticles like EVs that are very small in size and buoyant [10]. In the recent past advances in magnetic nanopartickes/beads based EVs enrichment and detection platforms have shown remarkable progress and are favoured for their simplicity, robustness and cost effectiveness [11]. Recently, Gao and fellows developed magnetic beads with titanium oxide coating (Fe_3O_4@TiO_2) for the enrichment of exosomes from serum. The enrichment of exosomes with Fe_3O_4@TiO_2 is assisted by strong bidentate binding between titanium oxide and phosphate group of the lipid bilayer of exosomes. In the complex biological systems for instance serum, the hydrophilic phosphate heads of phospholipids tend to be exposed on outer side of lipid bilayer. Fe_3O_4@TiO_2 based selective enrichment of exosomes is fast, efficient, and free of contamination due to non-specific adsorption [12]. The use of Fe_3O_4@TiO_2 beads greatly reduced the EVs' isolation time and reduced the loss of EVs during the enrichment step without affecting its proteome [13].

By targeting the surface markers of EVs many strategies have been developed for EVs detection such as colorimetric [14], fluorescence [15] and electrochemical [16]. The detection of EVs with tumor specific markers (such as PD-L1) is an interesting and worth exploring avenue of research in oncology [17]. Some methods reporting the detection of PD-L1@EVs have achieved good sensitivity, but are either based on sophisticated instruments such as flow cytometry [18] and droplet digital PCR [17] or are time consuming [19]. Many methods reporting the detection of PD-L1@EVs had used antibodies for PD-L1 recognition. However, antibodies have canonical limitations such as batch to batch variations, poor stability and high cost [14]. An alternative to antibodies is aptamer that can exert antibodies-mimic function. Aptamers are small size short oligonucleotides that can be synthesized synthetically to exhibit high specific binding affinity. They are stable, less pricey, amenable to chemical modifications and can be synthesized in large scale with controllable batch to batch variations, Therefore, aptamers are lucrative option for development of biosensing tools that can be applied in the physiological environment. Recently, CD63 specific aptamers [20] and PD-L1 specific aptamers [17] based detection platforms for EVs have shown excellent sensitivity and specificity.

As EVs are considered ideal biomarker for the diagnosis of variety of diseases, research on development of new detection platforms for EVs is rapidly expanding [21]. An important improvement in EVs detection methods is to enrich EVs directly from the serum without employing traditional EVs enrichment steps such as ultracentrifugation/ultrafiltration/expensive commercial kits or antibody coated magnetic beads [22]. We therefore, designed a chemiluminescent sensing assay for the detection of PD-L1@EVs by integrating $Fe_3O_4@TiO_2$ for rapid enrichment of EVs from a small volume of undiluted serum with PD-L1 aptamers specificity [23], and chemiluminescence sensitivity. Chemiluminescence is light emission due to chemical reaction, chemiluminescence based detection platforms have the potential to achieve lower detection limits and wide linear calibration range [24], therefore, can be valuable for clinical diagnosis [25] such as clinical diagnosis of cancer markers and EVs [26].

2. Materials and Methods

2.1. Aptamers Sequences and Auxiliary Reagents

The Biotin-PD-L1 aptamer, Cy3-PD-L1 aptamer, Scambled-PD-L1 aptamer were all synthesized by Sangon Biotech Co., Ltd. (Shanghai, China) The sequences are shown in Table S1. Streptavidin-ALP made by Bioss Antibodies (catalog number: bs-0437p-AP). L-Argnine, Sodium Chloride, and Magnesium chloride made by Macklin Biochemical Co., Ltd. (Shanghai, China) Tween-20 made by Beyotime. To rotate the tubes we used 3D Rotary made by MIULAB. Water purified by Mili-Q portable benchtop system was used throughout all the experiments and buffers preparation.

2.2. Cell Culturing for EVs Isolation

Human lung adenocarcinoma cells (A549) and Human lung epithelial cells (BEAS-2B) by Chinese Academy of Sciences were cultured (at 37 °C in 5% CO_2 and 95% air) in Dulbecco's modified Eagle medium (DMEM, Gibco) medium augmented with 10% Fetal Bovine Serum (FBS, Gibco). After achieving the 70% confluency, cells were washed with PBS then DMEM with 10% exosomes depleted FBS (System Bioscience) was poured to support the growth of the cells. After 24 h to 48 h duration, the media was collected to amass EVs via differential ultracentrifugation [27]. In short, culture media was exposed to centrifugation (Beckman Coulter, Allegra X-30R) at 300× *g* for 10 min, then 2000× *g* for 10 min, and 10,000× *g* for 30 min to mitigate unwanted components such as cells, dead cells, and cell debris, respectively. Upon completion of the above steps, the supernatant was filtered via 0.2 µm syringe cap filter then the filtered supernatant was ultracentrifuged (100,000× *g*) for 90 min (Beckman Coulter, Optima XPN-100). After heedful pouring of supernatant the pellet was resuspended in 1 mL PBS and exposed to ultracentrifugation (100,000× *g*) for 90 min using TYPE 90 Ti Rotor (Beckman). Finally, supernatant was poured out and EVs' pellet was re-suspended in 100 µL PBS (0.22 µm filtered) and stored at −80 °C to be used later in further experiments. All the centrifugation and ultracentrifugation steps were performed at 4 °C.

2.3. $Fe_3O_4@TiO_2$ Beads Formation

$Fe_3O_4@TiO_2$ beads were made following an established protocol [28,29]. The synthesis include the mixing of 0.05 g Fe_3O_4 with 0.3 mL aqueous Ammonia (28 wt%) in a 100 mL Ethanol solution. The amalgam was stirred (600 rpm) for 15 min at 45 °C in a three neck flask. Then about 0.75 mL Tetrabutyl titanate solution made in Ethanol was introduced drop-wise into the solution under continuous stirring for almost 24 h at 45 °C. Later in a Teflon-lined stainless-steel autoclave, the mixture was exposed to hydrothermal treatment at 180 °C (24 h). The final product was then washed with water and Ethanol (10 mL) and dried overnight. After drying the final product was stored for subsequent experiments.

2.4. EVs Morphology, Characterization and Enumeration

Transmission electron microscopy (TEM) was used for visualization of EVs morphology. To accomplish this pipetted 20 µL EVs over a copper mesh and the superficial liquid was mitigated via filter paper. Then about 2% phosphotungstic acid negative staining solution was applied for 5 min. The copper mesh was then heedfully washed (5X) with double distilled water and dried prior to visualization via JEM—2200CX TEM (JEOL). Protein concentration of the samples was measured using Qubit fluorometer (Invitrogen) for westren blotting (WB). EVs and cells were lysed by heating for 5 min at 95 °C in the loading buffer that were then loaded on 10% SDS-PAGE gel (Sangon Biotech) for separating the proteins. Polyvinylidene Difluoride (PVDF) membrane (Sangon Biotech) was selected for transferring the protein bands from gel to the membrane. After transferring, the membrane was blocked using 2% non-protein blocking solution for 2 h. The membrane was washed (3X) before dispersing in the solutions of primary antibodies (CD9, TSG101, Calnexin [ab275018 by Abcam] and PD-L1 antibody by Proteintech) and kept at 4 °C for 12 h. After incubation with primary antibodies the membrane was then washed (3X) to remove unbound primary antibodies and incubated for 2 h with solution of secondary antibody (goat anti-rabbit, Abcam) at room temperature. Finally, enhanced chemiluminescence kit (Sangon Biotech) was used to visualize the protein bands. To get the information about the concentration and size distribution of EVs employed an NTA technique (ZetaView, Particle Metrix, Diessen, Germany). NTA readings for each sample were measured at 11 different positions. The result were visualized using ZetaView® software. All the frames were automatically analyzed to drop out outlier positions and calculated the EVs concentration and mean size.

2.5. Confocal Microscopy and Zeta Potential

For confocal microscopy EVs bound $Fe_3O_4@TiO_2$ beads were blocked using blocking solution. The blocking solution was made by mixing 5% Bovine serum albumin (BSA, Sangon Biotech, Shanghai, China) and 0.5 mg/mL Salmon sperm DNA (Invitrogen) in HEPES buffer. After blocking added Cy3-PD-L1 aptamer and incubated for 30 min. The unbound aptamers were removed by washing (3X) and the beads + EVs + Cy3-PD-L1 aptamer were re-suspended in 300 μL PBS and examined using confocal microscope (Leica microsystems, 100X oil immersion objective). PBS was used as a control. For Zeta potential of $Fe_3O_4@TiO_2$ beads (0.1 mg/mL), A549 released EVs (10^6 EVs/mL) and $Fe_3O_4@TiO_2$+ EVs complex (0.1 mg/mL) used Zetasizer nanoseries (Nano-ZS).

2.6. EVs Accumulation by $Fe_3O_4@TiO_2$ Beads

To measure the EVs enrichment capability of the $Fe_3O_4@TiO_2$ beads, a PKH26 dye-stained A549 EVs were used [28]. PKH26 staining of A549 EVs was achieved following manufacture instructions. Samples containing equal concentration of PKH26-stained A549 EVs (10^{10} EVs/ mL) were made and recorded the fluorescence. The stained EVs were then incubated with $Fe_3O_4@TiO_2$ beads a magnetic rack was used to pellet $Fe_3O_4@TiO_2$/EVs and measured the supernatant fluorescence. The experiment was conducted in triplicate and the capture efficiency was calculated as capture efficiency(%) = (F2/F1) × 100; where F1 is the fluorescence of EVs solution before enrichment by $Fe_3O_4@TiO_2$ beads and F2 after enrichment by $Fe_3O_4@TiO_2$ beads. Likewise, optimal quantity of $Fe_3O_4@TiO_2$ beads and optimal time needed for achieving optimal capturing efficiency were also evaluated.

2.7. Clinical Feasibility

Clinical feasibility was elucidated using the serum samples of healthy individuals (n = 7) and LC patients (n = 18) . After acquiring written consent and prior ethical approval from the ethical committee, we obtained serum samples at the Clinical laboratory of the Drum tower hospital affiliated with the Nanjing Medical University. The serum was filtered using 0.22 μm filter fitted as a syringe cap. About 50 μL of the undiluted serum was incubated with the $Fe_3O_4@TiO_2$ beads for the duration of 10 min to capture EVs and followed the remaining steps of sensing assay to accomplish the detection. PBS was used as a blank control. For verification purpose all the samples were tested in triplicate.

2.8. Statistical Testing

To perform graphing and statistical operations used GraphPad prism software (Inc., La Jolla, CA, USA).

3. Results

3.1. Principal Design of the Sensing Assay

Figure 1 depicts the principal design of the sensing assay. First of all the undiluted serum was incubated with $Fe_3O_4@TiO_2$ for 10 min to capture EVs. After capturing EVs the beads were pelleted using magnetic force to remove the supernatant. To block the non-specific binding sites a blocking solution containing salmon sperm DNA (0.5 mg/mL) and BSA (3% Vol/Vol) in HEPES buffer (10 mM) was added and incubated for 1 h with gentle rotation. The beads were then pelleted to remove the blocking solution and washed (TBS, pH 7.5 + 0.001% Tween + 1 mM $MgCl_2$). The remaining sites were blocked by adding 500 mM Arginine in HEPES buffer for 30 min. After magnetic pelleting of beads and removal of supernatant Biotin-PD-L1 aptamer (0.075 µm in 3% BSA-HEPES, the binding buffer) was added and incubated for 30 min with gentle rotation. Magnetically pellet the beads to rinse the excessive aptamers followed by washing and addition of Streptavidin-ALP (1 mg/mL, dilution 1:3000) for about 20 min. Similarly, the beads were pelleted and the excessive Streptavidin-ALP was rinsed followed by the addition of CDP-Star: a chemiluminescent substrate of ALP and was gently rotated in the dark for 8–10 min. Finally, $Fe_3O_4@TiO_2$ beads were magnetically separated and the supernatant was pipetted into the 96-wells half area white-plates. The chemiluminescence was recorded at 466 nm using a spectrophotometer (SpectraMax M5). The whole sensing assay completes at room temperature in less than 3 h.

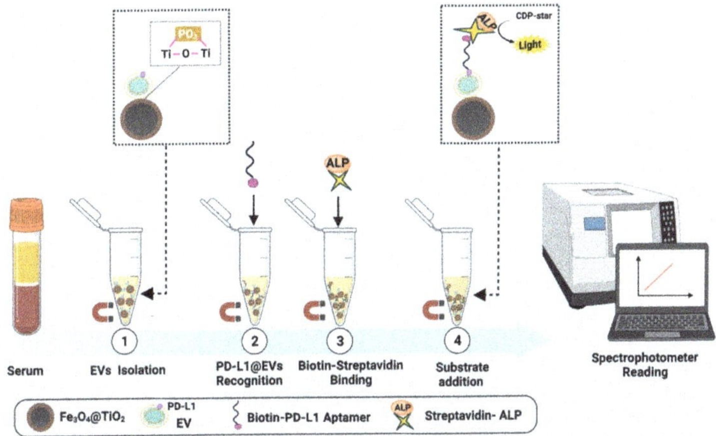

Figure 1. Principal design of the assay. (**1**) $Fe_3O_4@TiO_2$ based enrichment of EVs from undiluted serum. (**2**) Addition of Biotin-PD-L1 Aptamer to bind to PD-L1@EVs. (**3**) Addition of Streptavidin-Alakaline phosphatase (ALP) to achieve Biotin-Streptavidin binding. (**4**) Addition of CDP-Star for catalysis by ALP and subsequent reading of chemiluminescense by spectrophotometer.

3.2. $Fe_3O_4@TiO_2$ Formation Analysis

Formation of the $Fe_3O_4@TiO_2$ beads was analyzed using TEM (Figure S1A,B) and SEM (Figure S1C,D). The size of the beads was from 300 nm to 500 nm. The beads appear to have a black core of Fe_3O_4 and granulated surface coating of TiO_2. Energy-dispersive X-ray spectroscopy based elemental analysis of $Fe_3O_4@TiO_2$ beads confirmed the presence of Ti and Fe that also establish a successful coating of TiO_2 over Fe_3O_4 beads (Figure S2). $Fe_3O_4@TiO_2$ beads were then harnessed to capture EVs from PBS/serum based on TiO_2 binding affinity with the phosphate group of the EV's lipid bilayer [12]. The Zeta potential of the $Fe_3O_4@TiO_2$ beads showed considerable change upon binding the negatively charged EVs. Hence, the changes in Zeta potential of $Fe_3O_4@TiO_2$ beads can also transcribe the binding of EVs with beads (Figure S3).

3.3. EVs Characterization, Enumeration and PD-L1 Realization

The morphological characterization by TEM confirmed the typical saucer-shaped morphology of the acquired EVs (Figure 2A). WB results validated the expression of EVs specific markers and cancer specific marker, we obtained distinct bands for CD9, TSG101 and PDL-1 verifying the presence of the EVs specific markers and cancer specific marker, respectively, while calnexin as an EVs' negative marker (Figure 2B). NTA analysis showed that we had acquired around 2.5×10^{11} EVs/mL from the A549 cells with the mean size range of 145 nm. Likewise, from BEAS-2B cells acquired 4.5×10^{9} EVs/mL with 120 nm mean size (Figure 2C). NTA results verify the fact that cancer cell releases more EVs than normal cells.

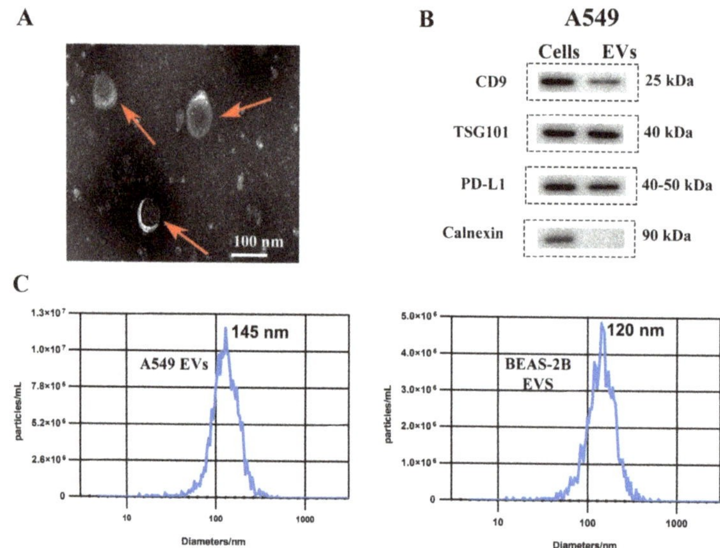

Figure 2. The characterization of canonical features of EVs. (**A**) TEM image of A549 released EVs red arrow points at EVs (**B**) western blot-based visualization of EVs' positive and negative markers and PD-L1 marker. (**C**) NTA graph showing EVs concentration and size distribution.

3.4. Assay's Conditions Optimization

In order to establish an excellent detection assay optimization of the key steps is crucial. One of the key step of the designed sensing assay is the capturing of EVs by $Fe_3O_4@TiO_2$ beads. We therefore, optimized the amount of beads and incubation time to achieve optimum capture efficiency of $Fe_3O_4@TiO_2$ beads for EVs. Different quantities of beads (0.2 mg, 0.4 mg, 0.6 mg, 0.8 mg and 1 mg) were tested to obtain the quantity of beads needed to achieve the highest capture efficiency for PKH26 labeled EVs. The florescence was recorded before and after the addition of $Fe_3O_4@TiO_2$ beads, optimum capture efficiency (more than 80%) was obtained for 0.6 mg $Fe_3O_4@TiO_2$ amount. Further increase in the amount showed no significant increment in capture efficiency, as shown in Figure 3A. Another important parameter was the optimization of incubation time needed for the $Fe_3O_4@TiO_2$ beads to achieve maximum capturing using PKH26 labeled EVs, various time duration from 2 min to 16 min were tested for optimal capture efficiency. Maximum capture efficiency was recorded at 10 min and no significant increment was achieved in capture efficiency post 10 min (Figure 3B).

PD-L1 aptamer has a central role in the design of this study, therefore, it was pertinent to evaluate its optimum concentration for optimal assay performance. Different concentrations of aptamers(0.025, 0.050, 0.075, 0.100, 0.500 µM) were tested to find the optimum concentration of aptamers. The concentration of EVs (10^{10} EVs/mL) was kept constant while PBS was used as blank. The best signal (sample CL intensity) to noise (PBS, CL intensity) was obtained for aptamer concentration of 0.075 µm as shown in Figure 3C. Similarly, to record best incubation time needed for the aptamer to produce good signal (CL intensity (a.u.) of EVs sample) as compared to the noise (CL intensity (a.u.) of the PBS). Three different incubation times (30 min, 60 min, 120 min) were trialed and best S/N ratio was recorded for 30 min incubation time. However, incubation times longer than 30 min resulted in increased background signals as shown in Figure 3D. To further study the specificity of the PD-L1 aptamer we compared it with scrambled PD-L1 aptamer and PBS (Figure S4). High CL intensity was obtained for PD-L1 aptamer as compared to scrambled aptamer and PBS. Validating PD-L1 aptamer high specificity.

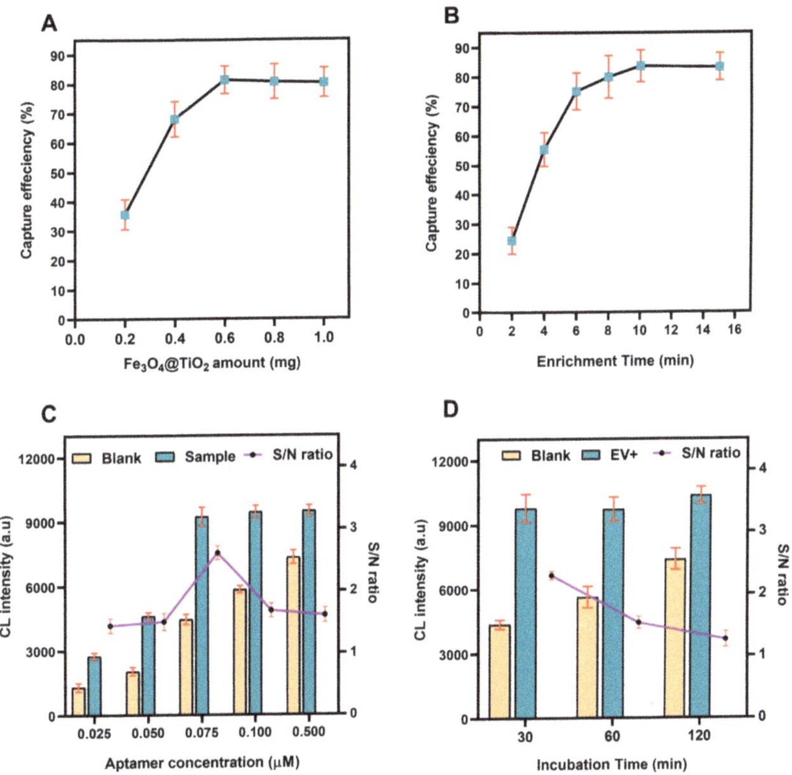

Figure 3. Optimization experiments. (**A**) $Fe_3O_4@TiO_2$ amount and optimal capture efficiency. (**B**) $Fe_3O_4@TiO_2$ incubation time and optimal capturing potential. (**C**) Aptamer concentrations for optimal performance of the assay. (**D**) Aptamer incubation time for optimal assay performance. (Error bar indicates the standard deviation of three replicates, S/N = signal-to-noise ratio).

3.5. Confocal Microscopy-Based Visualization of PDL@EVs

To verify the presence PD-L1@EVs we used confocal microscopy. Cy3-PD-L1 aptamer was combined with two combinations; (1) $Fe_3O_4@TiO_2$ with EVs and (2) $Fe_3O_4@TiO_2$ with PBS. Images obtained showed Cy3 specific fluorescence in the first combination while no fluorescence can be seen in the image of the second combination. From the images we can rightly conclude the capturing of EVs by $Fe_3O_4@TiO_2$ and can deduce the presence of PD-L1 over the surface of EVs as Cy3-PD-L1 aptamer can only bind to EVs expressing PD-L1 marker (Figure 4).

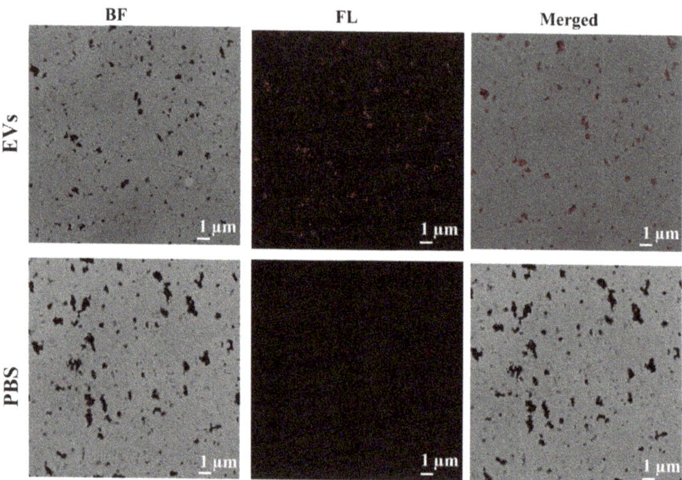

Figure 4. Confocal microscopy-based visualization of the PD-L1@EVs by Cy3-PD-L1 aptamer. BF (Bright filed), FL (Florescence).

3.6. Verification of Assay's Specificity, Sensitivity and Accuracy

Important features of an excellent assay are its specificity, sensitivity and accuracy. To test the specificity, the same concentration (10^8 EVs/mL) of cancer cell (A549) derived EVs and normal cells (BEAS-2B) derived EVs, FBS derived EVs, EVs derived from the serum of healthy individual were used. The results showed high CL intensity (a.u.) value for the A549 derived PD-L1@EVs as compared to the rest of the trailed EVs, establishing high specificity of sensing assay for PD-L1@EVs (Figure 5A).

Next, to assess the assay sensitivity different concentrations of the A549 released EVs were tested from 10^5–10^{10} EVs/mL. The results showed a direct relationship between the concentration of PD-L1@EVs and CL intensity (a.u.) value (Figure 5B). Increase in EVs concentration increase the CL intensity (a.u.). The calibration curve was drawn by plotting the values of CL intensity (a.u.) against the EVs concentration. Linear relationship existed between CL intensity and the EVs concentration from 10^5–10^8 EVs/mL. The LOD of the method was calculated to be 2.85×10^5 EVs/mL, using 3 times the standard deviation of the blank formula (Figure 5C).

To explicate the assay recovery ability, various concentrations of EVs (10^6, 10^7 and 10^8 EVs/mL) harvested from A549 cells were spiked into the EVs depleted healthy human serum. To calculate the percent (%) recovery; divided the detected EVs concentration by the concentration of EVs spiked into the serum multiplied by 100. It can be transpired from Table S2 that the assay has good recovery ($\geq$90%) for all the tested samples. For any biological assay, its ability to produce consistently good results is of utmost importance. We trailed the assay using the same number of EVs (10^7 EVs/mL) consistently for 5 times. The results were clustered around the mean value, with the relative standard deviation of 4.41% (Figure S5).

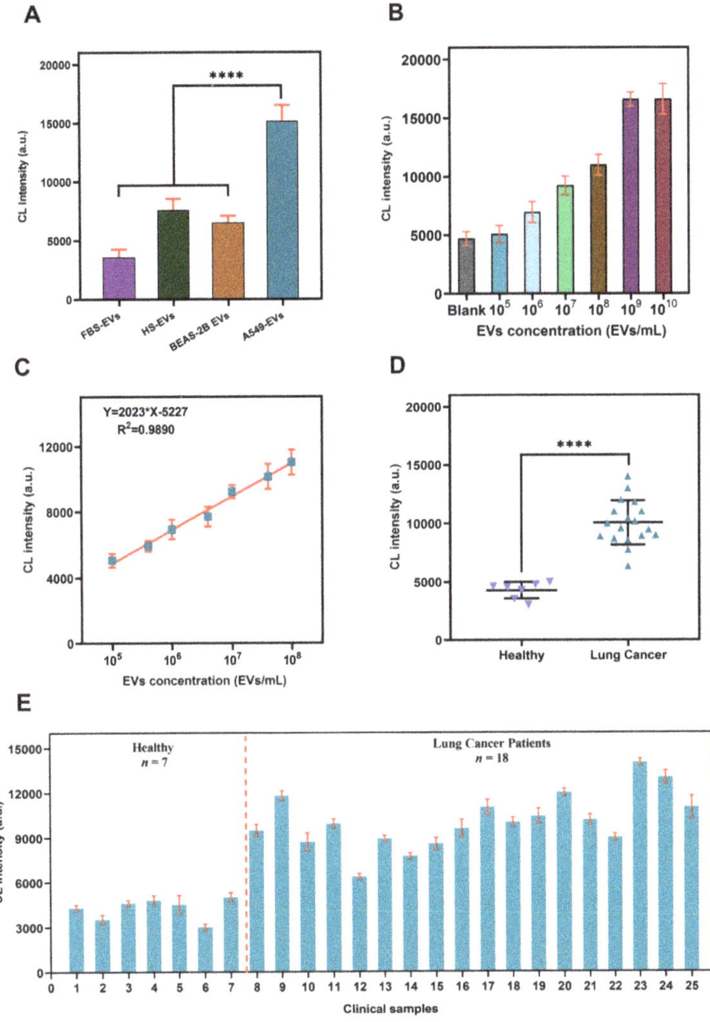

Figure 5. Assay features validation. (**A**) Assay specificity for PDL@EVs. HS (healthy serum) FBS (Fetal bovine serum). (**B**) Assay performance on different concentration of EVs. (**C**) Assay's linear fitting between CL intensity and concentration of EVs. (**D**,**E**) Assay performance validation using clinical serum samples. (Error bar indicates standard deviation of three replicates, **** = $p \leq 0.0001$).

3.7. Sensing PD-L1@EVs in Clinical Samples

An important feature of any detection method is its capability to perform well when tested using clinical serum samples; serum of healthy individuals (n = 7), and lung cancer patients (n = 18). The clinical information of the subjects used in the study are tabulated in Table S2. It is evident from Figure 5D,E that serum samples of healthy individuals have lower CL intensity value than lung cancer patients. The statistical significance of CL intensity (a.u.) value between two groups was analyzed using the Mann-Whitney U test ($p < 0.001$). These findings are in accordance with an already reported research that serum of LC patients have more PD-L1@EVs than the serum of healthy individuals [28]. To put it simple, the proposed assay is capable of differentiating between healthy and LC patients. Moreover, by taking into consideration the good performance of the sensing assay using clinical serum samples, it can be anticipated that the assay has good prospects for application in clinical setups for the detection of LC.

4. Discussion and Conclusions

Lung cancer (LC) rapid expansion in terms of new cases,its poor 5-year survival rate and towering mortality rate increased the demand for development of simple, sensitive, and non-invasive detection strategies [6]. EVs have been reported to be ideal for development of non-invasive diagnostic methods with excellent applicability [7]. PD-L1@EVs have been reported to be involved in immunosuppression and tumor immune escape [3]. Likewise, the quantity of PD-L1@EVs can be a biomarker for cancer and for evaluating the patients response to anti-PD-1/PD-L1 immunotherapy [5]. Taking into consideration the enormous potential of PD-L1@EVs as a biomarker [30], we designed a simple, specific and sensitive sensing assay for PD-L1@EVs. The salient features of the sensing assay are the $Fe_3O_4@TiO_2$ beads remarkable capability to capture EVs from the undiluted serum (50 µL), hence, eliminates the use of ultracentrifugation and other commercial methods for EVs isolation prior to detection. The use of aptamer instead of antibodies as a recognition element, have added advantages such as aptamer have excellent sensitivity, specificity and affinity towards the target and are less expensive. The use of CDP-star as an ALP substrate eliminates the need for H_2O_2 as a co-substrate thus minimizing the interference and cost. The assay exhibited good sensitivity with an LOD 2.54×10^5 EVs/mL and a wider linear range (10^5–10^8 EVs/mL). The performance of the assay remained significant in complex undiluted clinical serum samples establishing its utility for clinical samples. The sensing assay completes at room temperature without any expensive machine-based pretreatment.

The LOD and linear range of the sensing assay is better than rest of tabulated methods in Table 1. Most of the methods employed ultracentrifugation or other commercial kit prior to detection of EVs in clinical serum, some used simulated serum, and some used diluted serum with spiked EVs or just detected using cells culture medium. However, the proposed sensing assay showed good performance with real undiluted serum samples. Compared to the traditional ELISA assay for PD-L1@EVs reported before our sensing assay needs lesser time (less than 3 h). They have used costly EVs isolation kit, and the operation time is over 12 h [3]. Another method based on using the digital droplet PCR and aptamer for the detection of PD-L1@EVs had achieved good sensitivity. However the method need large sample volume and employed ultracentrifugation to enrich EVs prior to detection. The process of ultracentrifugation need an expensive instrument and is time consuming [3]. Zhang et al., achieved lower LOD but narrow liner range compared to the proposed sensing assay. They have also employed ultracentrifugation prior to detection of EVs that lengthens the detection operation [31]. Recently published research establish that PD-L1@EVs can be used for the diagnosis of LC [28] and breast cancer [32]. Our results also recognize the fact that the amount of PD-L1@EVs is distinct between healthy individuals and LC patients. PD-L1 has been reported to undergo heavy glycosylation blocking the binding of antibody; affecting the performance of antibody based analytical methods for PD-L1@EVs detection [33]. However, aptamers due to small size can overcome the PD-L1 glycosylation interference and can bind efficiently [5]. Thus, the use of aptamer in our study increases the sensitivity as well as decreases the cost compared to the antibody-based methods. To sum it up, we have developed a simple $Fe_3O_4@TiO_2$ and aptamer-based chemiluminescent sensing assay for EVs with PD-L1 marker for LC diagnosis. To conclude we have developed a simple, specific and sensitive chemiluminsent sensing assay for PD-L1@EVs by integrating $Fe_3O_4@TiO_2$ beads rapid enrichment of EVs from undiluted serum with PD-L1 aptamer specificity and cheminluminsence sensitivity. The assay has good prospects for application in clinical setups to classify between healthy and LC patients; it can also aide in the surveillance of patients' response to anti-PD-1/PD-L1 immunotherapy. In future we are looking to automate the assay for improving the sensitivity and linear range and to eliminate the loss of EVs due to differences in inter-personal pippeting skills during washing steps. In addition by using a large patients cohort data integrated with machine learning model for results analysis is recommended to further validate and improve the utility of our method for LC diagnosis. In order to expand the assay utility for surveillance of anti-PD-L1/PD-1 immunotherapy we are planning to collaborate with hospital to test the sensing assay using clinical serum samples of LC patients undergoing anti-PD-L1/PD-1 immunotherapy.

Table 1. Comparison with contemporary methods.

Method	LOD (EVs/mL)	Sample Type	Prior Isolation (UC/kit/UF)	Linear Range (EVs/mL)	Reference
Fluorescence	1.6×10^6	EVs Spiked Serum	Yes	1.66×10^6–1.6×10^8	[34]
Fluorescence	7.6×10^6	Culture Medium	Yes	1.68×10^7–4.2×10^{10}	[35]
FET	1.76×10^8	Diluted Serum	Yes	1.76×10^8–1.76×10^{10}	[36]
Optical	10^8	EVs Spiked Serum	Yes	3.9×10^7–2×10^{10}	[37]
Lateral flow immunoassay	8.54×10^8	Culture Medium	Yes	10^9–10^{11}	[38]
Colorimetric	5.2×10^8	Diluted Serum	Yes	1.84×10^9–2.2×10^{10}	[39]
Colorimetric	13.52×10^8	Diluted Serum	Yes	1.9×10^9–3.38×10^{10}	[14]
Chemiluminescence	2.63×10^8	Undiluted Serum	No	2.92×10^8–2.80×10^{11}	[26]
Chemiluminescence	2.85×10^5	Undiluted Serum	No	$10^5 - 10^8$	**This method**

Supplementary Materials: The following supporting information can be downloaded at: https://www.mdpi.com/article/10.3390/bios12050281/s1, Table S1. The aptamer sequences used in the sensing assay; Table S2. The recovery (%) of the assay; Table S3. Clinical information of the patients enrolled in the study; Figure S1. TEM and SEM images of Fe_3O_4@TiO_2 beads; Figure S2. XDS analysis of Fe_3O_4@TiO_2 beads; Figure S3. Zeta potential of Fe_3O_4@TiO_2 beads, EV and Fe_3O_4@TiO_2 beads with EVs. Figure S4. PD-L1 aptamer specificity; Figure S5. Repetitions for the confirming the analytical performance of the sensing assay.

Author Contributions: A.K. conceptualization; data curation; formal analysis and writing original draft. K.D. experimental support and discussion. H.K. writing—review and editing, experimental support and discussion. N.H. funding acquisition and resources. Z.L. funding acquisition, project administration, supervision, conceptualization, writing—review and editing. All authors have read and agreed to the published version of the manuscript.

Funding: This research was supported by funding from National Key Research and Development Program of China (2017YFA0205301), National Natural Science Foundation of China (No. 61971216, No. 81902153, and No. 62071119) and Key Research and Development Project of Jiangsu Province (No. BE2019603, BE2020768).

Institutional Review Board Statement: Study was approved by ethical review committee of the Drum Tower Hospital affiliated with the medical school of Nanjing University (No: 32000549, Approved: 22 April 2021.

Informed Consent Statement: Informed consent was obtained from all the subjects involved in the study.

Data Availability Statement: Data supporting the research findings are provided in the main manuscript and supporting information

Acknowledgments: We sincerely acknowledge the help of the following individuals throughout the research: Wang Rong, Irfan Tariq, Wasi Ullah Khan, Faisal Raza, Sifat Ullah Khan, Umer Sharif, Tauseef Ahmad and Xiaoshan Zhang.

Conflicts of Interest: No conflict of interest to be declared.

References

1. Kalluri, R.; LeBleu, V.S. The biology, function, and biomedical applications of exosomes. *Science* **2020**, *367*, eaau6977. [CrossRef] [PubMed]
2. Yekula, A.; Muralidharan, K.; Kang, K.M.; Wang, L.; Balaj, L.; Carter, B.S. From laboratory to clinic: Translation of extracellular vesicle based cancer biomarkers. *Methods* **2020**, *177*, 58–66. [CrossRef] [PubMed]

3. Chen, G.; Huang, A.C.; Zhang, W.; Zhang, G.; Wu, M.; Xu, W.; Yu, Z.; Yang, J.; Wang, B.; Sun, H.; et al. Exosomal PD-L1 contributes to immunosuppression and is associated with anti-PD-1 response. *Nature* **2018**, *560*, 382–386. [CrossRef]
4. Daassi, D.; Mahoney, K.M.; Freeman, G.J. The importance of exosomal PDL1 in tumour immune evasion. *Nat. Rev. Immunol.* **2020**, *20*, 209–215. [CrossRef] [PubMed]
5. Huang, M.; Yang, J.; Wang, T.; Song, J.; Xia, J.; Wu, L.; Wang, W.; Wu, Q.; Zhu, Z.; Song, Y.; et al. Homogeneous, low-volume, efficient, and sensitive quantitation of circulating exosomal PD-L1 for cancer diagnosis and immunotherapy response prediction. *Angew. Chem.* **2020**, *132*, 4830–4835. [CrossRef]
6. Sung, H.; Ferlay, J.; Siegel, R.L.; Laversanne, M.; Soerjomataram, I.; Jemal, A.; Bray, F. Global cancer statistics 2020: GLOBOCAN estimates of incidence and mortality worldwide for 36 cancers in 185 countries. *CA A Cancer J. Clin.* **2021**, *71*, 209–249. [CrossRef]
7. Liu, C.; Xiang, X.; Han, S.; Lim, H.Y.; Li, L.; Zhang, X.; Ma, Z.; Yang, L.; Guo, S.; Soo, R.; et al. Blood-based liquid biopsy: Insights into early detection and clinical management of lung cancer. *Cancer Lett.* **2022**, *524*, 91–102. [CrossRef]
8. Boriachek, K.; Masud, M.K.; Palma, C.; Phan, H.P.; Yamauchi, Y.; Hossain, M.S.A.; Nguyen, N.T.; Salomon, C.; Shiddiky, M.J. Avoiding pre-isolation step in exosome analysis: Direct isolation and sensitive detection of exosomes using gold-loaded nanoporous ferric oxide nanozymes. *Anal. Chem.* **2019**, *91*, 3827–3834. [CrossRef]
9. Holcar, M.; Kandušer, M.; Lenassi, M. Blood nanoparticles–influence on extracellular vesicle isolation and characterization. *Front. Pharmacol.* **2021**, 3178. [CrossRef]
10. Gu, Y.; Chen, C.; Mao, Z.; Bachman, H.; Becker, R.; Rufo, J.; Wang, Z.; Zhang, P.; Mai, J.; Yang, S.; et al. Acoustofluidic centrifuge for nanoparticle enrichment and separation. *Sci. Adv.* **2021**, *7*, eabc0467. [CrossRef]
11. Martín-Gracia, B.; Martín-Barreiro, A.; Cuestas-Ayllón, C.; Grazú, V.; Line, A.; Llorente, A.; Jesús, M.; Moros, M. Nanoparticle-based biosensors for detection of extracellular vesicles in liquid biopsies. *J. Mater. Chem. B* **2020**, *8*, 6710–6738. [CrossRef] [PubMed]
12. Gao, F.; Jiao, F.; Xia, C.; Zhao, Y.; Ying, W.; Xie, Y.; Guan, X.; Tao, M.; Zhang, Y.; Qin, W.; et al. A novel strategy for facile serum exosome isolation based on specific interactions between phospholipid bilayers and TiO_2. *Chem. Sci.* **2019**, *10*, 1579–1588. [CrossRef] [PubMed]
13. Zhang, N.; Sun, N.; Deng, C. Rapid isolation and proteome analysis of urinary exosome based on double interactions of Fe_3O_4@TiO_2-DNA aptamer. *Talanta* **2021**, *221*, 121571. [CrossRef] [PubMed]
14. Wang, Y.M.; Liu, J.W.; Adkins, G.B.; Shen, W.; Trinh, M.P.; Duan, L.Y.; Jiang, J.H.; Zhong, W. Enhancement of the intrinsic peroxidase-like activity of graphitic carbon nitride nanosheets by ssDNAs and its application for detection of exosomes. *Anal. Chem.* **2017**, *89*, 12327–12333. [CrossRef]
15. Yu, X.; He, L.; Pentok, M.; Yang, H.; Yang, Y.; Li, Z.; He, N.; Deng, Y.; Li, S.; Liu, T.; et al. An aptamer-based new method for competitive fluorescence detection of exosomes. *Nanoscale* **2019**, *11*, 15589–15595. [CrossRef]
16. Xie, H.; Di, K.; Huang, R.; Khan, A.; Xia, Y.; Xu, H.; Liu, C.; Tan, T.; Tian, X.; Shen, H.; et al. Extracellular vesicles based electrochemical biosensors for detection of cancer cells: A review. *Chin. Chem. Lett.* **2020**, *31*, 1737–1745. [CrossRef]
17. Lin, B.; Tian, T.; Lu, Y.; Liu, D.; Huang, M.; Zhu, L.; Zhu, Z.; Song, Y.; Yang, C. Tracing tumor-derived exosomal PD-L1 by dual-aptamer activated proximity-induced droplet digital PCR. *Angew. Chem. Int. Ed.* **2021**, *60*, 7582–7586. [CrossRef]
18. Gao, X.; Teng, X.; Dai, Y.; Li, J. Rolling circle Amplification-assisted flow cytometry approach for simultaneous profiling of exosomal surface proteins. *ACS Sens.* **2021**, *6*, 3611–3620. [CrossRef]
19. He, Y.; Wu, Y.; Wang, Y.; Wang, X.; Xing, S.; Li, H.; Guo, S.; Yu, X.; Dai, S.; Zhang, G.; et al. Applying CRISPR/Cas13 to construct exosomal PD-L1 ultrasensitive biosensors for dynamic monitoring of tumor progression in immunotherapy. *Adv. Ther.* **2020**, *3*, 2000093. [CrossRef]
20. Xu, L.; Chopdat, R.; Li, D.; Al-Jamal, K.T. Development of a simple, sensitive and selective colorimetric aptasensor for the detection of cancer-derived exosomes. *Biosens. Bioelectron.* **2020**, *169*, 112576. [CrossRef]
21. Wang, S.; Khan, A.; Huang, R.; Ye, S.; Di, K.; Xiong, T.; Li, Z. Recent advances in single extracellular vesicle detection methods. *Biosens. Bioelectron.* **2020**, *154*, 112056. [CrossRef] [PubMed]
22. Jiang, Q.; Liu, Y.; Wang, L.; Adkins, G.B.; Zhong, W. Rapid enrichment and detection of extracellular vesicles enabled by CuS-enclosed microgels. *Anal. Chem.* **2019**, *91*, 15951–15958. [CrossRef] [PubMed]
23. Hu, H.; Ding, Y.; Gao, Z.; Li, H. S1 nuclease digestion-based rational truncation of PD-L1 aptamer and establishment of a signal dual amplification aptasensor. *Sens. Actuators Chem.* **2021**, *331*, 129442. [CrossRef]
24. McPherson, R.A.; Pincus, M.R. *Henry's Clinical Diagnosis and Management by Laboratory Methods E-Book*; Elsevier Health Sciences: Philadelphia, PA, USA, 2021.
25. Worsfold, P.; Townshend, A.; Poole, C.F.; Miró, M. *Encyclopedia of Analytical Science*; Elsevier: Amsterdam, The Netherlands, 2019.
26. Wang, Y.; Liu, Z.; Wang, X.; Dai, Y.; Li, X.; Gao, S.; Yu, P.; Lin, Q.; Fan, Z.; Ping, Y.; et al. Rapid and quantitative analysis of exosomes by a chemiluminescence immunoassay using superparamagnetic iron oxide particles. *J. Biomed. Nanotechnol.* **2019**, *15*, 1792–1800. [CrossRef]
27. Thierry, C.; Amigorena, S.; Raposo, G.; Clayton, A. Isolation and characterization of exosomes from cell culture supernatants. *Curr. Protoc. Cell Biol.* **2006**, *3*, 1–29.
28. Pang, Y.; Shi, J.; Yang, X.; Wang, C.; Sun, Z.; Xiao, R. Personalized detection of circling exosomal PD-L1 based on Fe_3O_4@TiO_2 isolation and SERS immunoassay. *Biosens. Bioelectron.* **2020**, *148*, 111800. [CrossRef]

29. Li, W.; Yang, J.; Wu, Z.; Wang, J.; Li, B.; Feng, S.; Deng, Y.; Zhang, F.; Zhao, D. A versatile kinetics-controlled coating method to construct uniform porous TiO2 shells for multifunctional core–shell structures. *J. Am. Chem. Soc.* **2012**, *134*, 11864–11867. [CrossRef]
30. Dong, H.Y.; Xie, Q.H.; Pang, D.W.; Chen, G.; Zhang, Z.L. Precise selection of aptamers targeting PD-L1 positive small extracellular vesicles on magnetic chips. *Chem. Commun.* **2021**, *57*, 3555–3558. [CrossRef]
31. Zhang, H.; Zhou, Y.; Luo, D.; Liu, J.; Yang, E.; Yang, G.; Feng, G.; Chen, Q.; Wu, L. Immunoassay-aptasensor for the determination of tumor-derived exosomes based on the combination of magnetic nanoparticles and hybridization chain reaction. *RSC Adv.* **2021**, *11*, 4983–4990. [CrossRef]
32. Cao, Y.; Wang, Y.; Yu, X.; Jiang, X.; Li, G.; Zhao, J. Identification of programmed death ligand-1 positive exosomes in breast cancer based on DNA amplification-responsive metal-organic frameworks. *Biosens. Bioelectron.* **2020**, *166*, 112452. [CrossRef]
33. Lee, H.H.; Wang, Y.N.; Xia, W.; Chen, C.H.; Rau, K.M.; Ye, L.; Wei, Y.; Chou, C.K.; Wang, S.C.; Yan, M.; et al. Removal of N-linked glycosylation enhances PD-L1 detection and predicts anti-PD-1/PD-L1 therapeutic efficacy. *Cancer Cell* **2019**, *36*, 168–178. [CrossRef] [PubMed]
34. Pan, Y.; Wang, L.; Deng, Y.; Wang, M.; Peng, Y.; Yang, J.; Li, G. A simple and sensitive method for exosome detection based on steric hindrance-controlled signal amplification. *Chem. Commun.* **2020**, *56*, 13768–13771. [CrossRef] [PubMed]
35. Wang, L.; Yang, Y.; Liu, Y.; Ning, L.; Xiang, Y.; Li, G. Bridging exosome and liposome through zirconium–phosphate coordination chemistry: A new method for exosome detection. *Chem. Commun.* **2019**, *55*, 2708–2711. [CrossRef] [PubMed]
36. Kwong Hong Tsang, D.; Lieberthal, T.J.; Watts, C.; Dunlop, I.E.; Ramadan, S.; del Rio Hernandez, A.E.; Klein, N. Chemically functionalised graphene FET biosensor for the label-free sensing of exosomes. *Sci. Rep.* **2019**, *9*, 13946. [CrossRef] [PubMed]
37. Moura, S.L.; Martín, C.G.; Martí, M.; Pividori, M.I. Multiplex detection and characterization of breast cancer exosomes by magneto-actuated immunoassay. *Talanta* **2020**, *211*, 120657. [CrossRef]
38. Oliveira-Rodríguez, M.; López-Cobo, S.; Reyburn, H.T.; Costa-García, A.; López-Martín, S.; Yáñez-Mó, M.; Cernuda-Morollón, E.; Paschen, A.; Valés-Gómez, M.; Blanco-López, M.C. Development of a rapid lateral flow immunoassay test for detection of exosomes previously enriched from cell culture medium and body fluids. *J. Extracell. Vesicles* **2016**, *5*, 31803. [CrossRef]
39. Xia, Y.; Liu, M.; Wang, L.; Yan, A.; He, W.; Chen, M.; Lan, J.; Xu, J.; Guan, L.; Chen, J. A visible and colorimetric aptasensor based on DNA-capped single-walled carbon nanotubes for detection of exosomes. *Biosens. Bioelectron.* **2017**, *92*, 8–15. [CrossRef]

 *biosensors*

Communication

A Novel Enzyme-Free Ratiometric Fluorescence Immunoassay Based on Silver Nanoparticles for the Detection of Dibutyl Phthalate from Environmental Waters

Hui Meng [1,†], Nannan Yao [1,†], Kun Zeng [1], Nuanfei Zhu [1], Yue Wang [1], Biying Zhao [2] and Zhen Zhang [1,*]

[1] School of the Environment and Safety Engineering, Jiangsu University, Zhenjiang 212013, China; menghui@ujs.edu.cn (H.M.); 2211909008@stmail.ujs.edu.cn (N.Y.); kjj80116@ujs.edu.cn (K.Z.); zhunuanfei@163.com (N.Z.); wangyue_persnl@126.com (Y.W.)

[2] International Genome Center, Jiangsu University, Zhenjiang 212013, China; zhaoby@ujs.edu.cn

* Correspondence: zhangzhen@ujs.edu.cn; Fax: +86-511-88790955

† These authors contributed equally to this work.

Abstract: A novel ratiometric fluorescent immunoassay was developed based on silver nanoparticles (AgNPs) for the sensitive determination of dibutyl phthalate (DBP). In the detection system, AgNPs were labeled on the secondary antibody (AgNPs@Ab_2) for signal amplification, which aimed to regulate the H_2O_2 concentrations. When AgNPs-Ab_2 and antigen–primary antibody (Ab_1) were linked by specific recognition, the blue fluorescence of Scopoletin (SC) could be effectively quenched by the H_2O_2 added while the red fluorescence of Amplex Red (AR) was generated. Under the optimized conditions, the calculated detection of limit (LOD, 90% inhibition) reached 0.86 ng/mL with a wide linear range of 2.31–66.84 ng/mL, which was approximately eleven times lower than that by HRP-based traditional ELISA with the same antibody. Meanwhile, it could improve the inherent built-in rectification to the environment by the combination of the dual-output ratiometric fluorescence assays with ELISA, which also enhanced the accuracy and precision (recoveries, 87.20–106.62%; CV, 2.57–6.54%), indicating it can be applied to investigate the concentration of DBP in water samples.

Keywords: immunoassay; antibody; dibutyl phthalate; scopoletin; amplex red; silver nanoparticles

Citation: Meng, H.; Yao, N.; Zeng, K.; Zhu, N.; Wang, Y.; Zhao, B.; Zhang, Z. A Novel Enzyme-Free Ratiometric Fluorescence Immunoassay Based on Silver Nanoparticles for the Detection of Dibutyl Phthalate from Environmental Waters. *Biosensors* **2022**, *12*, 125. https://doi.org/10.3390/bios12020125

Received: 16 January 2022
Accepted: 15 February 2022
Published: 16 February 2022

1. Introduction

Dibutyl phthalate (DBP), a type of phthalate ester (PAE), is widely applied in the industry, medicine, agriculture, and domestication [1–3]. As it does not react chemically with the molecular bonds of polymer matrix, DBP can be easily released into the environment and has become a ubiquitous environmental contaminant. Previous studies have indicated that DBP can interfere with the normal hormone secretion of the human body, and has reproductive, embryological, and genetic toxicity even at trace level [4,5]. Considering its biological accumulation and slow biodegradation in the organization [1], developing efficient methods to detect DBP for environmental risk assessment is urgently needed.

Nowadays, instrumental methods are commonly used to detect DBP [6–8]. These methods suffer from high cost, low throughput, and tedious sample pretreatment not to mention requiring advanced instruments and professional operation. Hence, these approaches are unsuitable for field studies and on-line monitoring of DBP [6]. By contrast, the enzyme-linked immunosorbent assay (ELISA) has become the most widely used analysis techniques in many fields, such as diseases diagnosis [9,10], food safety [11,12], and environmental monitoring [13,14]. It shows obvious advantages owing to its low cost, good specificity, automation, and high-throughput [11]. However, the traditional colorimetric ELISA which mainly relies on natural enzymes biocatalyzing a chromogenic substrate to colored molecules as signal output often suffers from low signal outputs, formidable background, and thus relatively low sensitivity and selectivity [15]. Considering that, a more sensitive and accurate immunosensor is urgently needed to fulfill the requirement for DBP monitoring.

Currently, many strategies, such as chemiluminescent ELISA [16], plasmonic ELISA [17], electrochemical ELISA [18] and fluorescence ELISA [19], have been developed to overcome the main drawbacks of the traditional colorimetric ELISA. Among this research, ratiometric fluorescence ELISA is a promising strategy for monitoring DBP in complex environmental matrices due to its superior properties such as high sensitivity and satisfactory accuracy, which are based on the ratio of the intensities of two different emission peaks as a signal [20,21] that not only reduces external interference but also improves detection sensitivity through self-calibration [22]. To improve the sensitivity and selectivity of the ratiometric fluorescence sensor, many attempts have been made using nanoparticles as an alternative to the nature enzyme or fluorescent labels. Zhu et al. [23] designed a novel ratiometric fluorescence ELISA based on a catalytic oxidation of OPD using as-prepared MnO_2NF-mediated signal tagging with the help of exogenous CDs for C-reactive protein (CRP) detection. Lin et al. [24] developed a ratiometric fluorescence enzyme-linked immunosorbent assay for sensitive detection of ethyl carbamate (EC) by introducing fluorescent silicon nanoparticles (SiNPs) into the chromogenic substrate system (o-phenylenediamine (OPD)/H_2O_2). Although these excellent works showed that ratiometric fluorescence sensors have a huge potential for trace detection, the advantage of enzyme-mediated signal amplification, which is important for high-throughput analysis, was not applied [25–27].

In this study, a ratiometric fluorescence ELISA to detect DBP was developed. The principle of the sensing method is shown in Scheme 1. Firstly, silver nanoparticles (AgNPs) were prepared by a one-step method and applied in the ratiometric fluorescence ELISA system, which displayed excellent peroxidase-like activity toward the substrate Scopoletin (SC) and Amplex Red (AR). In addition, AgNPs were labeled on the secondary antibody (AgNPs@Ab_2) to adjust the concentrations of H_2O_2, which was related to the magnitude of the signal response. After being optimized and evaluated, the proposed high-throughput method was applied to investigate the occurrence of DBP in a typical area.

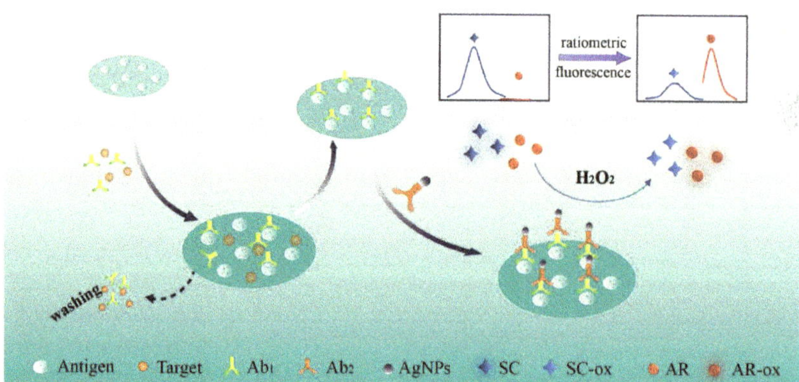

Scheme 1. Schematic illustration of a simple ratiometric fluorescence ELISA for TBBPA determination based on AgNPs and H_2O_2.

2. Experiments and Materials

2.1. Synthesis of Ag Nanoparticles (AgNPs)

All the glass instruments used in the following procedure were repeatedly washed with ultra-pure water and dried in an oven at 37 °C. AgNPs were prepared by the published method with slight modifications [28]. Briefly, $AgNO_3$ (0.1 g) was dissolved in 200 mL of deionized water and the solution transferred to a round-bottomed flask. When the solution was heated to a boil, 1% of sodium citrate (10 mL) was added slowly and heating was continued. AgNPs were synthesized successfully until the solution turned pale yellow. The obtained solution was stirred at 13,000 rpm for 15 min and the supernatant was also dialyzed for 3 days. Finally, the prepared solution was stored at 4 °C before use.

2.2. Synthesis of AgNPs@Ab_2

The preparation of antibody marker AgNPs@Ab_2 was based on the synthesis method of Zhang et al. [29], and the specific process was as follows: 1 mL of AgNP solution was put into a vial, to which 10 μL of sheep anti-rabbit solution (2 mg/mL) was added. The mixture was stirred at 37 °C for 12 h, then 500 μL 1.0 wt% BSA solution was added and stirring continued at a constant temperature. The resulting products were centrifuged, shaken, washed, redissolved in PBS solution, and stored at 4 °C for future use.

2.3. Catalytic Oxidation Assay

The catalytic activities of synthesized AgNPs were evaluated by fluorescence changes of SC and AR in the presence of H_2O_2. The specific experiment was carried out as follows: 200 μL of H_2O_2 solution (125 mM) was mixed with 10 μL of SC or AR in the absence or presence of AgNPs in a 500 μL centrifuge tube at 37 °C for 30 min. To ensure that SC and AR did not interfere with each other in the homogeneous system, we conducted the following experiments to verify: 200 μL of H_2O_2 solution (125 mM) was added into 20 μL of SC and AR with AgNPs or not in a 500 μL centrifuge tube at 37 °C for 30 min.

2.4. The Development of Ratiometric Fluorescence ELISA for DBP

First, DBP antigen (100 μL/well) was coated on the 96-well microplate at 4 °C using coating buffer overnight. After washing four times with PBS, each well was blocked with blocking buffer at 37 °C for 45 min. Then, 50 μL DBP antibody and 50 μL DBP were added to each well separately and kept at 37 °C for 30 min. Washing again, 100 μL of AgNPs-labeled secondary antibody was added into each well. After the steps of incubation and washing, 200 μL H_2O_2 and 20 μL SC and AR were added into each well at 37 °C for 30 min. Finally, the fluorescence intensity was measured by a microplate reader under the excitation wavelengths of 380 nm and 560 nm.

2.5. LC–MS/MS Analysis

To verify the reliability of this established method, samples of DBP in real water were determined by the LC–MS/MS method. The experiment details are described in the Supplementary Materials.

3. Result and Discussion

3.1. The Characterization of AgNPs and AgNPs@Ab_2

Figure 1C shows that the ultraviolet absorption spectrum of bar AgNPs has a strong peak defined maximum close to 400 nm, consistent with the literature [30]. As shown in Figure 1A,B, we can see that AgNPs have a structure with a roughly spherical shape and a size of about 30 nm, which increased to 50 nm upon conjugation with Ab_2. This increase was due to the agglomeration of particles that is often caused by the bonding between Ab_2 and AgNPs [31]. As seen from Figure 1C, the ultraviolet absorption peak of AgNPs@Ab_2 was obviously blue-shifted, indicating the successfully conjugations of AgNPs and Ab_2 [29]. To further evaluate the synthesized AgNPs@Ab_2, the Zeta potential values of AgNPs and AgNPs@Ab_2 were measured to be −15.77 mV and −4.74 mV, respectively (Figure 1D). AgNPs@Ab_2 was significantly lower than AgNPs, which was due to secondary antibody and silver nanoparticles being bound to each other by electrostatic adsorption, disrupting the homeostasis of the original colloid [32]. The result of the Zeta potential also proved the successful synthesis of AgNPs laterally.

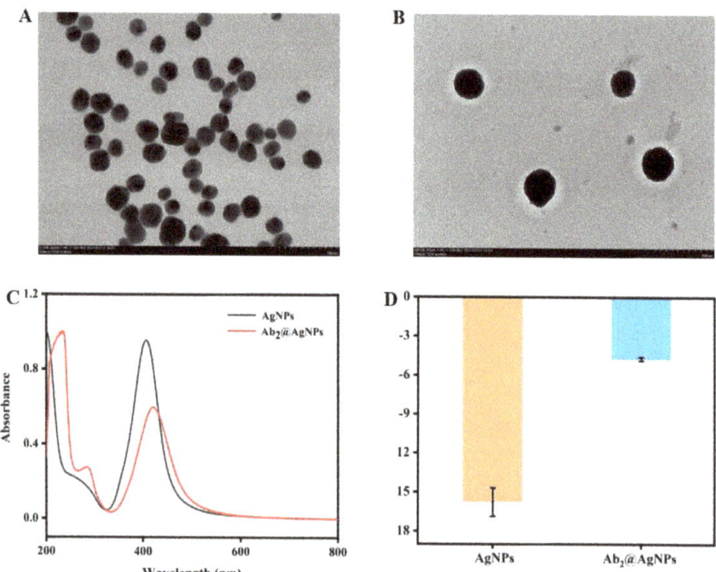

Figure 1. Characterization of AgNPs and Ab_2@AgNPs. (**A**) Structure and TEM image of AgNPs, (**B**) structure and TEM image of Ab_2@AgNPs, (**C**) The UV absorption spectra of AgNPs and AgNPs@Ab_2, and (**D**) The Zeta potential analysis of AgNPs and AgNPs@Ab_2.

3.2. The Peroxidase-like Activity of the Nanoparticles

As an important noble metal nanoparticle, AgNP has attracted extensive attention due to its low cost and high extinction coefficient [33]. Exploring the peroxidase-like properties of AgNPs can expand the application field of simulating the activity of nanoparticles [34,35]. To demonstrate the peroxidase activity of AgNPs, we tested the catalytic oxidation of SC and AR in the presence of H_2O_2, two fluorescent substrates that possess inverse responses. The emission peak of SC was at 465 nm, and produced almost non-fluorescent products (FI_{465}, SC_{-OX}) by oxidation. In contrast, AR was non-fluorescent, but could produce a significantly enhanced signal at 585 nm (FI_{585}, AR_{-OX}) after H_2O_2's oxidation under the catalysis of peroxidase [36]. As shown in Figure 2A, SC presented a high fluorescent signal in the absence of H_2O_2 (curve a), but its fluorescence plummeted after the addition of AgNPs (curve b); AR had almost no fluorescence (curve c), while fluorescence was obviously enhanced at 585 nm (curve d). To ensure that Ag^+ and H_2O_2 had no effect on substrate fluorescence, we verified this by measuring fluorescence at different concentrations of Ag^+ and H_2O_2. Figure 3A,B indicates that Ag^+ and H_2O_2 had no effect on the fluorescence signal. This was further illustrated that the change of fluorescence was caused by AgNPs. We also investigated whether the two substrates interfere with each other in a homogeneous system. As seen from Figure 2B, the 3D fluorescence columns of the two substrates in the absence or presence of AgNPs performed exactly the same as in their respective uses. These results proved that AgNPs have good oxidase-like properties.

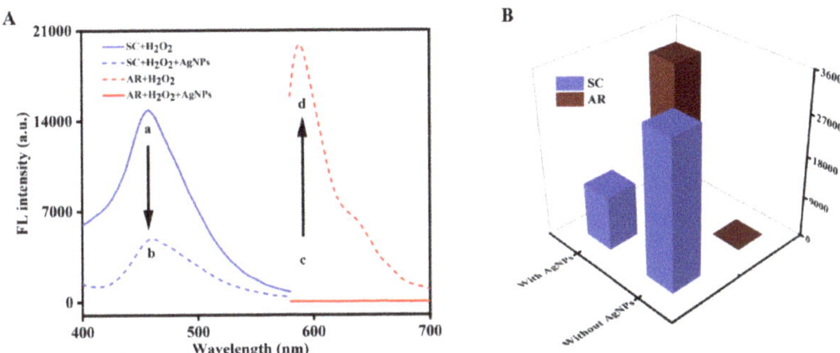

Figure 2. (**A**) Fluorescence spectra of separately used SC in (a) the absence of AgNPs, (b) the presence of AgNPs, and (c,d) separately used AR. (**B**) 3D fluorescent columns of simultaneously used SC (blue columns) and AR (red columns) without and with AgNPs.

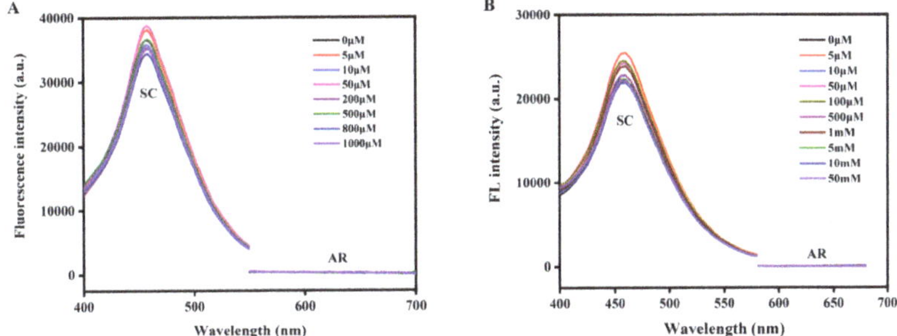

Figure 3. Fluorescence spectra of SC and AR under the different concentrations of Ag^+ (**A**) and H_2O_2 (**B**).

3.3. Optimization of Experimental Conditions

In order to obtain better sensing performance, several of the parameters of immunoassay were optimized, including the concentrations of H_2O_2, AR, and SC, the reaction time of AgNPs@Ab_2 and H_2O_2, and the pH. The fluorescence intensity ratio (F_{585}/F_{465}) was an important indicator for evaluating the performance of fluorescent immunoassay. Figure S1C indicates that with increasing H_2O_2, the fluorescence intensity ratio increased significantly until the concentration reached 125 mM. This was because 125 mM H_2O_2 can completely etch AgNPs, and beyond this level more H_2O_2 would not participate in the reaction. Therefore, 125 mM was selected as the optimal reaction concentration of H_2O_2 to achieve the best analytical performance. Additionally, Figure S1B also demonstrates that the most suitable reaction time between H_2O_2 and AgNPs was 30 min. Figure S1A reveals that with increasing of the pH value, F_{585}/F_{465} initially increased and then decreased at pH 7, which was selected as an optimal value in the assay. At the same time, Figure S1D,E show that the optimal concentrations of AR and SC were 20 μM and 4 μM, respectively.

Under the above optimized parameters, the fluorescence intensity of SC at 465 nm gradually decreased with the increasing of DBP concentration, while that of AR at 585 nm gradually increased (Figure 4A,B), which indicated that the fluorescence intensity ratio (F_{585}/F_{465}) was negatively correlated with the DBP concentration. As shown in Figure 4C, the standard curve was drawn with Origin 9.0 based on several different concentrations of DBP standard solutions and the fluorescence intensity ratio (F_{585}/F_{465}). In addition, the linear regression equation of $(F_{585}/F_{465})/(F_{585}/F_{465})_0 = 0.01388 + 0.98997/\{1 + (x/12.42809)^{0.82405}\}$ was obtained. Based on the curve, the limit of detection (LOD, 90% inhibition) was calculated as 0.86 ng/mL, which was almost 11 times lower than that of the conventional ELISA using the same antibody (seen in Figure 4D).

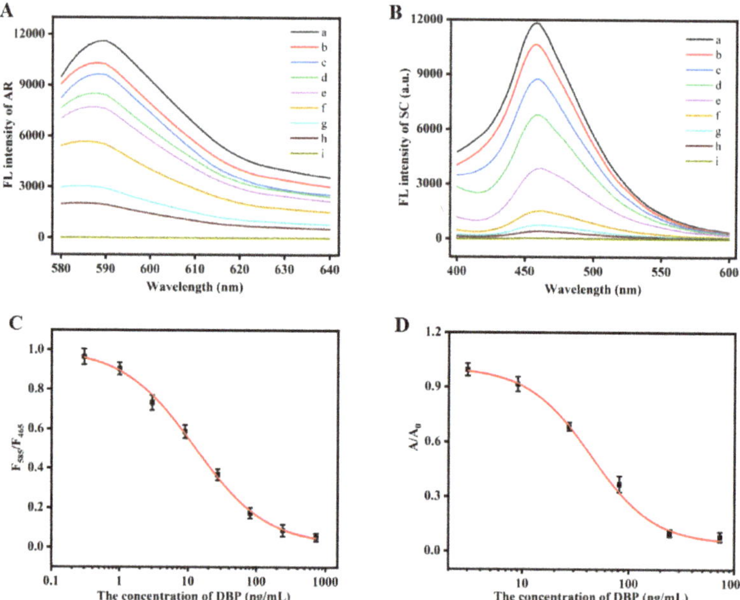

Figure 4. Fluorescence spectra of AR (**A**) and Sc (**B**) in the presence of increasing DBP concentrations. (**C**) Calibration curve of ratiometric fluorescence values (F_{465}/F_{585}) versus different concentrations of DBP (The error bars are obtained via three replicates) and (**D**) the standard curve of traditional indirect competitive ELISA for DBP under optimized conditions.

3.4. Assay Validation and Real Sample Analysis

The established method was estimated for its accuracy and precision through a spike-recovery analysis of water samples from various sources (pure water, pond water, tap water, and river water) fortified with a variety of DBP concentrations. As shown in Table 1, the satisfactory recoveries of DBP (87.2–106.62%) were obtained with the intra-assay coefficient of variation (CV) ranging from 2.57% to 6.54% by measuring three replicates, indicating that the developed method had good accuracy and could be applied for the detection of DBP from environmental samples.

Table 1. Analysis of DBP-spiked samples using this method (n = 3).

Samples	Background (ng/mL)	Added (ng/mL)	Found (ng/mL)	Recovery (%)	CV [a] (%)
Pure water	ND [b]	5	4.36	87.20	3.73
		15	15.31	102.07	2.57
		60	58.95	98.25	4.85
River water	10.46	5	16.14	104.40	4.97
		15	23.78	93.40	2.71
		60	71.42	101.36	3.68
Tap water	ND [b]	5	4.83	96.60	6.54
		15	13.67	91.13	5.23
		60	59.25	98.75	3.93
Pond water	21.71	5	25.97	97.23	2.91
		15	39.14	106.62	4.75
		60	75.37	92.24	3.46

CV [a]: The intra-assay coefficient of variation obtained from three determinations; ND [b]: not detected.

The proposed method was applied to detect potential DBP in real water samples from the Zhenjiang region, together with traditional ELISA and instrument methods to further verify its reliability. As shown in Tables 2 and S1, both results had a relatively high degree of agreement with the developed method. Meanwhile, our ratiometric fluorescence immunoassay showed better sensitivity than conventional ELISA.

Table 2. Comparison of DBP detection using our established method and the conventional ELISA ($n = 3$).

Samples	Background (ng/mL)	ELISA (ng/mL)	Our Method (ng/mL)	CV [a] (%)
S1	ND [b]	ND	1.12	3.57
S2	ND	ND	ND	ND
S3	ND	ND	3.97	2.35
S4	ND	ND	ND	ND
S5	ND	23.81	24.15	7.31
S6	ND	16.57	19.62	4.68
S7	ND	ND	ND	ND
S8	ND	ND	ND	ND
S9	ND	11.21	10.53	2.14
S10	ND	ND	ND	ND
S11	ND	9.62	8.36	3.95
S12	ND	19.87	20.65	2.93
S13	ND	ND	ND	ND
S14	ND	ND	ND	ND
S15	ND	ND	ND	ND

CV [a]: Intra-assay coefficient of variation obtained from 3 determinations performed in same polystyrene microtiter plate; ND [b]: not detected.

4. Conclusions

In summary, a ratiometric fluorescence immunoassay was established for high-throughput determination of DBP based on H_2O_2 etching of AgNPs. Reactive oxygen species (·OH) could be produced during etching, which played a key role in fluorescence quenching of SC and fluorescence generation of AR. Under the optimized conditions, the developed method showed good accuracy and reproducibility (recoveries, 87.20–106.62%; CV, 2.57–6.54%) with higher sensitivity (LOD, 0.86 ng/mL) than traditional ELISA. Further, we believe that this work could serve as a general platform for the detection of other substances with different antibodies.

Supplementary Materials: The following supporting information can be downloaded at: https://www.mdpi.com/article/10.3390/bios12020125/s1, Materials and Instruments. LC-MS/MS analysis. Figure S1. The optimization of the concentration of H_2O_2 (A), SC (B), and AR (C); the optimization of different pH (D); and the optimization of different reaction times (E). Table S1. Comparison of the results of LC–MS/MS and our method using randomly spiked samples.

Author Contributions: Data curation, N.Y.; formal analysis, N.Y.; funding acquisition, Z.Z.; methodology, N.Y.; supervision, H.M. and Z.Z.; validation, K.Z. and N.Z.; visualization, Y.W.; writing—original draft, N.Y.; writing—review and editing, H.M., B.Z. and Z.Z. All authors have read and agreed to the published version of the manuscript.

Funding: The present work was supported by the National Natural Science Foundation of China (Grants 21876067, 21577051, and 31800386), the Chinese Postdoctoral Science Fund (No. 2019M651721) and the Jiangsu Collaborative Innovation Center of Technology and Material of Water Treatment.

Institutional Review Board Statement: Not applicable.

Informed Consent Statement: Not applicable.

Data Availability Statement: All data are contained within the article.

Acknowledgments: The author would like to thank Zhenjiang Zhuanbo Testing Technology Co., Ltd. (Zhenjiang, China) for providing the JEM-2100 equipment.

Conflicts of Interest: The authors declare no conflict of interest. They have no commercial interest or any kind of association that might pose a conflict of interest with any entity or form.

References

1. Gao, Y.; An, T.; Ji, Y.; Li, G.; Zhao, C. Eco-toxicity and human estrogenic exposure risks from OH-initiated photochemical transformation of four phthalates in water: A computational study. *Environ. Pollut.* **2015**, *206*, 510–517. [CrossRef]
2. Montevecchi, G.; Masino, F.; Zanasi, L.; Antonelli, A. Determination of phthalate esters in distillates by ultrasound-vortex-assisted dispersive liquid-liquid micro-extraction (USVADLLME) coupled with gas chromatography/mass spectrometry. *Food Chem.* **2017**, *221*, 1354–1360. [CrossRef]
3. Wang, M.; Yang, F.; Liu, L.; Cheng, C.; Yang, Y. Ionic Liquid-Based Surfactant Extraction Coupled with Magnetic Dispersive μ-Solid Phase Extraction for the Determination of Phthalate Esters in Packaging Milk Samples by HPLC. *Food Anal. Methods* **2016**, *10*, 1745–1754. [CrossRef]
4. Zhang, J.; Jin, S.; Zhao, J.; Li, H. Effect of dibutyl phthalate on expression of connexin 43 and testosterone production of leydig cells in adult rats. *Environ. Toxicol. Pharmacol.* **2016**, *47*, 131–135. [CrossRef]
5. Janjua, N.R.; Mortensen, G.K.; Andersson, A.M.; Kongshoj, B.; Wulf, H.C. Systemic Uptake of Diethyl Phthalate, Dibutyl Phthalate, and Butyl Paraben Following Whole-Body Topical Application and Reproductive and Thyroid Hormone Levels in Humans. *Environ. Sci. Technol.* **2007**, *41*, 5564–5570. [CrossRef]
6. Zhang, Z.; Zeng, K.; Liu, J. Immunochemical detection of emerging organic contaminants in environmental waters. *TrAC Trends Anal. Chem.* **2017**, *87*, 49–57. [CrossRef]
7. Han, X.; Liu, D. Detection of the toxic substance dibutyl phthalate in Antarctic krill. *Antarct. Sci.* **2017**, *29*, 511–516. [CrossRef]
8. Ibrahim, N.; Osman, R.; Abdullah, A.; Saim, N. Determination of Phthalate Plasticisers in Palm Oil Using Online Solid Phase Extraction-Liquid Chromatography (SPE-LC). *J. Chem.* **2014**, *2014*, 682975. [CrossRef]
9. Iha, K.; Inada, M.; Kawada, N.; Nakaishi, K.; Watabe, S.; Tan, Y.H.; Shen, C.; Ke, L.Y.; Yoshimura, T.; Ito, E. Ultrasensitive ELISA Developed for Diagnosis. *Diagnostics* **2019**, *9*, 78. [CrossRef]
10. Hsu, Y.-P.; Yang, H.-W.; Li, N.-S.; Chen, Y.T.; Pang, H.-H.; Pang, S.T. Instrument-Free Detection of FXYD3 Using Vial-Based Immunosensor for Earlier and Faster Urothelial Carcinoma Diagnosis. *ACS Sens.* **2020**, *5*, 928–935. [CrossRef]
11. Xiong, Y.; Leng, Y.; Li, X.; Huang, X.; Xiong, Y. Emerging strategies to enhance the sensitivity of competitive ELISA for detection of chemical contaminants in food samples. *TrAC Trends Anal. Chem.* **2020**, *126*, 115861. [CrossRef]
12. Bao, K.; Liu, X.; Xu, Q.; Su, B.; Liu, Z.; Cao, H.; Chen, Q. Nanobody multimerization strategy to enhance the sensitivity of competitive ELISA for detection of ochratoxin A in coffee samples. *Food Control* **2021**, *127*, 108167. [CrossRef]
13. Liu, Z.; Zhang, B.; Sun, J.; Yi, Y.; Li, M.; Du, D.; Zhu, F.; Luan, J. Highly efficient detection of salbutamol in environmental water samples by an enzyme immunoassay. *Sci. Total Environ.* **2018**, *613*, 861–865. [CrossRef]
14. Pan, Y.; Wei, X.; Guo, X.; Wang, H.; Song, H.; Pan, C.; Xu, N. Immunoassay based on Au-Ag bimetallic nanoclusters for colorimetric/fluorescent double biosensing of dicofol. *Biosens. Bioelectron.* **2021**, *194*, 113611. [CrossRef]
15. Liu, Y.; Pan, M.; Wang, W.; Jiang, Q.; Wang, F.; Pang, D.-W.; Liu, X. Plasmonic and Photothermal Immunoassay via Enzyme-Triggered Crystal Growth on Gold Nanostars. *Anal. Chem.* **2019**, *91*, 2086–2092. [CrossRef]
16. Vdovenko, M.M.; Stepanova, A.S.; Eremin, S.A.; Van Cuong, N.; Uskova, N.A.; Yu Sakharov, I. Quantification of 2,4-dichlorophenoxyacetic acid in oranges and mandarins by chemiluminescent ELISA. *Food Chem.* **2013**, *141*, 865–868. [CrossRef]
17. Xuan, Z.; Li, M.; Rong, P.; Wang, W.; Li, Y.; Liu, D. Plasmonic ELISA based on the controlled growth of silver nanoparticles. *Nanoscale* **2016**, *8*, 17271–17277. [CrossRef]
18. Pang, Y.-H.; Guo, L.-L.; Shen, X.-F.; Yang, N.-C.; Yang, C. Rolling circle amplified DNAzyme followed with covalent organic frameworks: Cascade signal amplification of electrochemical ELISA for alfatoxin M1 sensing. *Electrochim. Acta* **2020**, *341*, 136055. [CrossRef]
19. Wu, Y.; Guo, W.; Peng, W.; Zhao, Q.; Piao, J.; Zhang, B.; Wu, X.; Wang, H.; Gong, X.; Chang, J. Enhanced Fluorescence ELISA Based on HAT Triggering Fluorescence "Turn-on" with Enzyme–Antibody Dual Labeled AuNP Probes for Ultrasensitive Detection of AFP and HBsAg. *ACS Appl. Mater. Interfaces* **2017**, *9*, 9369–9377. [CrossRef]
20. Mao, G.; Cai, Q.; Wang, F.; Luo, C.; Ji, X.; He, Z. One-Step Synthesis of Rox-DNA Functionalized CdZnTeS Quantum Dots for the Visual Detection of Hydrogen Peroxide and Blood Glucose. *Anal. Chem.* **2017**, *89*, 11628–11635. [CrossRef]
21. Wang, J.; Jiang, C.; Jin, J.; Huang, L.; Yu, W.; Su, B.; Hu, J. Ratiometric Fluorescent Lateral Flow Immunoassay for Point-of-Care Testing of Acute Myocardial Infarction. *Angew. Chem. Int. Ed. Engl.* **2021**, *60*, 13042–13049. [CrossRef]
22. Zhu, N.; Li, X.; Liu, Y.; Liu, J.; Wang, Y.; Wu, X.; Zhang, Z. Dual amplified ratiometric fluorescence ELISA based on G-quadruplex/hemin DNAzyme using tetrahedral DNA nanostructure as scaffold for ultrasensitive detection of dibutyl phthalate in aquatic system. *Sci. Total Environ.* **2021**, *784*, 147212. [CrossRef]
23. Jiao, L.; Zhang, L.; Du, W.; Li, H.; Yang, D.; Zhu, C. Hierarchical manganese dioxide nanoflowers enable accurate ratiometric fluorescence enzyme-linked immunosorbent assay. *Nanoscale* **2018**, *10*, 21893–21897. [CrossRef]
24. Luo, L.; Song, Y.; Zhu, C.; Fu, S.; Shi, Q.; Sun, Y.-M.; Jia, B.; Du, D.; Xu, Z.-L.; Lin, Y. Fluorescent silicon nanoparticles-based ratiometric fluorescence immunoassay for sensitive detection of ethyl carbamate in red wine. *Sens. Actuators B Chem.* **2018**, *255*, 2742–2749. [CrossRef]

25. Zhao, J.; Wang, S.; Lu, S.; Liu, G.; Sun, J.; Yang, X. Fluorometric and Colorimetric Dual-Readout Immunoassay Based on an Alkaline Phosphatase-Triggered Reaction. *Anal. Chem.* **2019**, *91*, 7828–7834. [CrossRef]
26. Fan, Y.; Lv, M.; Xue, Y.; Li, J.; Wang, E. In Situ Fluorogenic Reaction Generated via Ascorbic Acid for the Construction of Universal Sensing Platform. *Anal. Chem.* **2021**, *93*, 6873–6880. [CrossRef]
27. Zhao, D.; Li, J.; Peng, C.; Zhu, S.; Sun, J.; Yang, X. Fluorescence Immunoassay Based on the Alkaline Phosphatase Triggered in Situ Fluorogenic Reaction of o-Phenylenediamine and Ascorbic Acid. *Anal. Chem.* **2019**, *91*, 2978–2984. [CrossRef]
28. Zhao, L.J.; Yu, R.J.; Ma, W.; Han, H.X.; Tian, H.; Qian, R.C.; Long, Y.T. Sensitive detection of protein biomarkers using silver nanoparticles enhanced immunofluorescence assay. *Theranostics* **2017**, *7*, 876–883. [CrossRef]
29. Zhang, Z.; Zhu, N.; Zou, Y.; Wu, X.; Qu, G.; Shi, J. A novel, enzyme-linked immunosorbent assay based on the catalysis of AuNCs@BSA-induced signal amplification for the detection of dibutyl phthalate. *Talanta* **2018**, *179*, 64–69. [CrossRef]
30. Schiesaro, I.; Battocchio, C.; Venditti, I.; Prosposito, P.; Burratti, L.; Centomo, P.; Meneghini, C. Structural characterization of 3d metal adsorbed AgNPs. *Phys. E Low-Dimens. Syst. Nanostruct.* **2020**, *123*, 114162. [CrossRef]
31. Tran, L.; Park, S. Highly sensitive detection of dengue biomarker using streptavidin-conjugated quantum dots. *Sci. Rep.* **2021**, *11*, 15196. [CrossRef]
32. Alvandi, H.; Dorosti, N.; Afshar, F. Synthesis of AgNPs and Ag@MoS_2 nanocomposites by dracocephalum kotschyi aqueous extract and their antiacetylcholinesterase activities. *Mater. Technol.* **2021**, 1–12. [CrossRef]
33. Du, P.; Niu, Q.; Chen, J.; Chen, Y.; Zhao, J.; Lu, X. "Switch-On" Fluorescence Detection of Glucose with High Specificity and Sensitivity Based on Silver Nanoparticles Supported on Porphyrin Metal–Organic Frameworks. *Anal. Chem.* **2020**, *92*, 7980–7986. [CrossRef]
34. Jiang, Z.J.; Liu, C.Y.; Sun, L.W. Catalytic Properties of Silver Nanoparticles Supported on Silica Spheres. *J. Phys. Chem. B* **2005**, *109*, 1730–1735. [CrossRef]
35. Jiang, C.; Bai, Z.; Yuan, F.; Ruan, Z.; Wang, W. A colorimetric sensor based on Glutathione-AgNPs as peroxidase mimetics for the sensitive detection of Thiamine (Vitamin B1). *Spectrochim. Acta Part A Mol. Biomol. Spectrosc.* **2022**, *265*, 120348. [CrossRef]
36. Fan, D.; Shang, C.; Gu, W.; Wang, E.; Dong, S. Introducing Ratiometric Fluorescence to MnO2 Nanosheet-Based Biosensing: A Simple, Label-Free Ratiometric Fluorescent Sensor Programmed by Cascade Logic Circuit for Ultrasensitive GSH Detection. *ACS Appl. Mater. Interfaces* **2017**, *9*, 25870–25877. [CrossRef]

Article

Bacterial Lighthouses—Real-Time Detection of *Yersinia enterocolitica* by Quorum Sensing

Julia Niehues [1], Christopher McElroy [1], Alexander Croon [1], Jan Pietschmann [1], Martin Frettlöh [2] and Florian Schröper [1,*]

[1] Fraunhofer Institute for Molecular Biology and Applied Ecology IME, Forckenbeckstraße 6, 52074 Aachen, Germany; julia.niehues@ime.fraunhofer.de (J.N.); christopher.mcelroy@ime.fraunhofer.de (C.M.); alexander.croon@ime.fraunhofer.de (A.C.); jan.pietschmann@ime.fraunhofer.de (J.P.)
[2] Quh-Lab Lebensmittelsicherheit, Siegener Str. 29, 57080 Siegen, Germany; martin.frettloeh@quh-lab.de
* Correspondence: florian.schroeper@ime.fraunhofer.de; Tel.: +49-(0)241-6085-13012

Abstract: Foodborne zoonotic pathogens have a severe impact on food safety. The demand for animal-based food products (meat, milk, and eggs) is increasing, and therefore faster methods are necessary to detect infected animals or contaminated food before products enter the market. However, conventional detection is based on time-consuming microbial cultivation methods. Here, the establishment of a quorum sensing-based method for detection of foodborne pathogens as *Yersinia enterocolitica* in a co-cultivation approach using a bacterial biosensor carrying a special sensor plasmid is described. We combined selective enrichment with the simultaneous detection of pathogens by recording autoinducer-1-induced bioluminescent response of the biosensor. This new approach enables real-time detection with a calculated sensitivity of one initial cell in a sample after 15.3 h of co-cultivation, while higher levels of initial contamination can be detected within less than half of the time. Our new method is substantially faster than conventional microbial cultivation and should be transferrable to other zoonotic foodborne pathogens. As we could demonstrate, quorum sensing is a promising platform for the development of sensitive assays in the area of food quality, safety, and hygiene.

Keywords: autoinducer; co-cultivation bioassay; food safety; foodborne pathogens; *N*-acyl homoserine lactones; plasmid-based biosensor; *Yersinia enterocolitica*

Citation: Niehues, J.; McElroy, C.; Croon, A.; Pietschmann, J.; Frettlöh, M.; Schröper, F. Bacterial Lighthouses—Real-Time Detection of *Yersinia enterocolitica* by Quorum Sensing. *Biosensors* **2021**, *11*, 517. https://doi.org/10.3390/bios11120517

Received: 5 November 2021
Accepted: 11 December 2021
Published: 16 December 2021

1. Introduction

In 1959, the World Health Organization (WHO) defined zoonoses as diseases that are naturally transmittable between humans and other vertebrates [1]. Up to 60% of human infectious diseases are thought to have a zoonotic origin [2]. Due to the growth of the human population, the demand for food products such as meat, milk and eggs, and the intensification of production is increasing. By this, the risk of zoonotic pathogens contaminating food and causing severe health effects is also increasing [3]. Campylobacteriosis was the most common zoonosis reported in 2019 with approximately 221,000 cases in the European Union, followed by salmonellosis (88,000 cases), infections with shiga toxin-producing *Escherichia coli* (STEC) (8000 cases), and yersiniosis (7000 cases) [4]. In 99% of yersiniosis cases, the agent was *Yersinia enterocolitica* [5].

The genus *Yersinia* (family *Enterobacteriaceae*) comprises 19 species, including the human and animal pathogens *Y. enterocolitica*, *Y. pestis*, and *Y. pseudotuberculosis* [6]. *Y. enterocolitica* is a Gram-negative bacillus, with a coccoid to rod-shaped morphology. This psychrophilic species can grow at temperatures within the range 0–45 °C, with optimal growth at 25–28 °C. *Y. enterocolitica* is motile at 25 °C, but loses motility at 37 °C [7,8]. The heterogeneous strains of *Y. enterocolitica* are assigned to six biotypes on the basis of their biochemical features (1A, 1B, and 2–5) and comprise more than 70 serotypes depending on their O-lipopolysaccharide determinants [9,10]. Generally, biotype 1B strains are highly virulent, whereas biotypes 2–5 have comparably low virulence. Strains of biotype 1A are generally non-virulent. Virulence is conferred by the plasmid pVY and virulence-associated chromosomal genes [11]. Nevertheless, some biotype 1A strains cause yersiniosis even though they lack these virulence components [6].

Y. enterocolitica is usually transferred to humans via the consumption of contaminated raw or undercooked pork, or by direct contact with contaminated carcasses [12]. Yersiniosis is an acute form of gastroenteritis that often includes symptoms such as fever and intestinal inflammation with watery or occasionally bloody diarrhea [7]. Pigs are the main reservoir, carrying this pathogen in their tonsils, lymphatic tissues, and intestine. As infected pigs are asymptomatic, the bacteria usually were not detected until the infected tissue is exposed during slaughter and processing, allowing the pathogen to spread over the surface of the carcasses [12,13]. Especially due to the devastating health effects, a regular testing for zoonotic pathogens is extremely important. However, rapid and reliable (molecular) detection methods for analysis of meat are not yet available [13].

The conventional detection of *Y. enterocolitica* requires microbial cultivation. An isolation step is necessary, consisting of selective enrichment followed by plating onto selective agar medium. Presumptive colonies are identified by biochemical and pathogenic phenotype. Depending on the choice of enrichment strategy, isolation may take from 48 h up to several weeks [14–16]. The conventional method is therefore time-consuming, laborious, and makes it difficult to detect low numbers of pathogens in the presence of abundant background flora also capable of growing on selective media [16].

The polymerase chain reaction (PCR) allows the genotypic characterization of *Y. enterocolitica* by amplifying the virulence-associated genes. Multiplex PCRs have been developed, and more sensitive detection can be achieved by real-time PCR [6,12,13,17–19]. PCR is useful for preliminary screening because it enables fast and sensitive analysis, but it requires expensive equipment and skilled personnel. Establishing reliable assays is not trivial since the presence of inhibitors in untreated samples often leads to false-negative results. Additionally, false-positive results may arise due to the detection of dead cells [15]. Immuno-based methods such as the enzyme-linked immunosorbent assay (ELISA) or lateral flow assays for whole cell detection are rare. ELISA formats detecting *Y. enterocolitica* directly are usually serotype-specific and are not well suited for sensitive detection of a bacterial infection [9]. Indirect diagnostic assays were developed to detect antibodies against *Y. enterocolitica* in animal blood and fecal samples [20,21]. However, this only allows the detection of *Y. enterocolitica*-specific antibodies, which only permits a conclusion about (a possible past) yersiniosis. Hence, these methods are not applicable for the detection of an actual bacterial contamination, e.g., in food samples. By this, current commercially available ELISA-based methods are not feasible for efficient analysis and detection of food contamination with *Y. enterocolitica*.

To address the limitations of conventional detection methods and current molecular assays, we developed a new approach for the rapid and sensitive detection of bacteria using quorum sensing (QS) signaling molecules. QS is generally known as a bacterial cell-cell communication process to coordinate gene expression depending on population density. In contrast to low cell density (LCD) populations, at which individual gene expression is favored, the synchronous expression of multiple genes is beneficial at high cell density (HCD) populations. QS-controlled processes include antibiotic production, sporulation, bioluminescence, biofilm formation, and virulence factor secretion. The QS system is based on signaling molecules called autoinducers (AIs), which are produced and secreted by cells. After reaching a local threshold concentration at HCD, the AIs trigger a signal cascade that modulates gene expression [22,23].

The AIs are divided into three major groups: AI-1 (AI-1), AI-2 (AI-2), and peptide-based AIs (AIPs) [24]. AIPs are a heterogeneous group of modified oligopeptides used by Gram-positive bacteria mainly for intraspecies communication [25]. AI-2 is a group of 4,5-dihydroxy-2,3-pentanedione (DPD) derivatives that can rapidly cyclize into furanone counterparts. AI-2 compounds are produced by a wide range of Gram-positive and Gram-negative bacteria and can be used to communicate between cells of the same or different species [26]. AI-1 compounds are *N*-Acyl homoserine lactone-based molecules (AHLs), consisting of a homoserine lactone ring moiety and an acyl side chain, used for intraspecies communication in Gram-negative bacteria [27]. AHL specificity is determined by the length, substitution, and saturation of the acyl side chain [28]. *Y. enterocolitica* produces the AHLs *N*-hexanoyl-L-homoserine lactone (HHL) and *N*-(3-oxohexanoyl)-L-homoserine lactone (OHHL) [29]. Therefore, we focused on the detection of AI-1 in this study.

The best-characterized QS system as a paradigm of intraspecies communication among Gram-negative bacteria is the LuxI/R system of the bioluminescent bacterium *Vibrio fischeri*. The generation of bioluminescence is regulated by the LuxI and LuxR proteins. LuxI is an AHL synthase that produces OHHL. The *luxI* gene is normally expressed at a basal level, resulting in the production of small quantities of OHHL [30]. AHLs are amphipathic molecules that can pass the cell membrane by diffusion [28]. As the cell population grows, the intracellular and extracellular concentration of AHL increases. When it reaches a local threshold concentration, the cytoplasmic LuxR protein binds to OHHL, and the resulting complex functions as a transcriptional activator, inducing the expression of *luxI* and the *lux* operon (*luxCDABEG*). LuxI therefore initiates a positive feedback loop by synthesizing the ligand that binds to LuxR and induces the synthesis of more LuxI. The increase in OHHL levels also triggers the expression of the *lux* operon in adjacent cells [30,31]. The structural genes of the *lux* operon generate bioluminescence via a luciferase-catalyzed oxidation reaction [32].

Within the last decades, the application of bacterial biosensor techniques has advanced the study of AHLs (AI-1) in QS systems and enables a simple method for qualitative and quantitative detection of AHLs [33–35]. The bacterial biosensor, which does not produce AHLs by itself, carries a sensor plasmid that encodes the *luxR* gene or the gene of a LuxR homologue, the cognate promoter, as well as a reporter gene for the detection. The presence of a compatible AHL finally induces the expression of the reporter gene whose product is detectable by, for example, bioluminescence [33]. In a previous study, several sensor plasmids (pSB401, pSB403, pSB406, pSB1075) were constructed by linking *luxR* together with the promoter *luxI'*, or homologues as *lasR* and *rhlR* with corresponding promoters, to the *lux* operon (*luxCDABE*) of *Photorhabdus luminescens* [36]. This *lux* operon offers a non-destructive real-time detection as a reporter gene and a higher temperature stability than the *lux* operon of *V. fischeri*, which is relevant for expression in recombinant *E. coli* at 37 °C [36,37]. Additionally, the induction of bioluminescence formation by activating AHLs was demonstrated [36].

In contrast, physico-chemical methods such as GC-MS or LC-MS/MS are frequently used for identification and quantification of AHLs due to their high specificity. However, these devices are highly expensive and need highly trained staff for handling, running, and maintaining the equipment. Furthermore, for method development highly trained staff are required, and sensitivity can vary greatly depending on method parameters, machine settings, and sample preparation. As demonstrated by Bauer and colleagues (2016), GC-MS methods, e.g., showed LODs of just 3.2 to 6.2 μM of single AHLs [38]. Targeted LC–MS/MS methods showed LODs of 0.23 to 1.51 nM [39]. Untargeted LC-MS/MS methods that are capable of detecting and identifying known and unknown AHL entities of a sample showed LODs of 0.6 to 229.6 nM [40]. Whole cell Biosensor approaches, as recently reviewed by Miller and Gilmore (2020), reach higher sensitivities, typically in the low nM and pM ranges [35].

To overcome the limitations of conventional cultivation methods, we established a new whole-cell biosensor approach that enables faster, more sensitive, and more reliable detection of *Y. enterocolitica* based on QS. Therefore, we used the *V. fischeri* LuxI/R system for detection of *Y. enterocolitica* by monitoring the production of its AHLs using a bacterial biosensor carrying a modified sensor plasmid derived from pSB401 [36]. The presence of OHHL in a sample induces the *lux* operon by binding the LuxR protein, inducing bioluminescence and indicating whether the sample is contaminated with *Y. enterocolitica* (Figure 1).

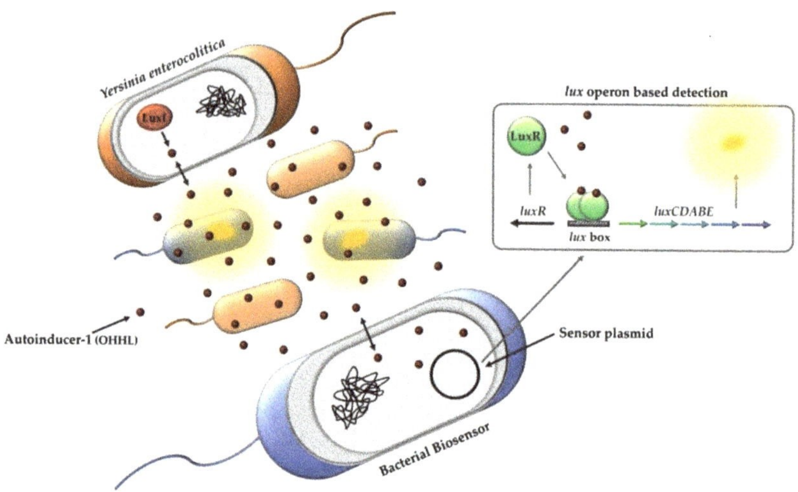

Figure 1. Schematic illustration of whole cell biosensor approach for *Y. enterocolitica* detection based on QS. OHHL produced and secreted by *Y. enterocolitica* induces the *lux* operon in recombinant *E. coli* biosensor resulting in luminescent signal response.

We first investigated the functionality of the biosensor in the presence of synthetic OHHL. We proved the ability to recognize AHLs produced by *Y. enterocolitica* and simultaneously tracked the AHL production of this pathogen via LC-MS/MS analysis for comparison. Finally, biosensor and pathogen were combined in an assay approach to achieve the highest sensitivity possible.

2. Materials and Methods

2.1. Sensor Plasmid pMA-RQ_luxR_lux

The customized sensor plasmid pMA-RQ_luxR_lux was synthesized by GeneArt (Thermo Fisher Scientific, Waltham, MA, USA). It was based on the sequence of pSB401 [36], which encodes a fusion construct combining the *Vibrio fischeri luxRI'* and *Photorhabdus luminescens luxCDABE* sequences. The fusion construct was joined to the pMA-RQ backbone, which contains an ampicillin-resistance gene and a ColE1 origin of replication (Figure S1).

2.2. Generation of the pMA Biosensor

The pMA-RQ_luxR-lux sensor plasmid was transformed into chemically competent *Escherichia coli* strain BL21 (DE3) cells (New England Biolabs, Frankfurt am Main, Germany) according to the manufacturer's protocol. Aliquots of 50 µL and 100 µL and the pellet were plated on lysogeny broth (LB) selection plates (10 g L^{-1} tryptone, 5 g L^{-1} yeast extract, 10 g L^{-1} NaCl, 15 g L^{-1} agar; pH 7.4) containing 100 µg mL^{-1} ampicillin. After an overnight incubation at 37 °C, successfully transformed cells were isolated on further selection plates and incubated overnight (37 °C). A colony was then used for preparation of fresh overnight cultures of the biosensor.

2.3. Luminescence Bioassay Procedure

An overnight culture in 5 mL selective LB was prepared with a clone from the master plate and incubated overnight at 37 °C, shaking at 160 rpm. On the next day, 25 mL of selective LB broth was inoculated with the overnight culture at a 1:1000 ratio. The biosensor culture was cultivated at 37 °C, shaking at 160 rpm, until the culture reached OD_{600} = 0.4. The culture was directly used for the luminescence assay. We mixed 20 µL of sample (synthetic AI standard concentrations or sterile culture supernatants) with 180 µL of the biosensor culture in C8 LockWell Lumi white 96-well plates (Thermo Fisher Scientific, Massachusetts, USA). For background measurement, the corresponding buffer or medium was used as blank. Kinetic measurements were carried out for 5 h at 30 °C recording luminescence signals every 20 min using a CLARIOstar plate reader (BMG Labtech, Ortenberg, Germany) with the following set parameters: top optic, 11.0 mm focal height, 10 s measuring interval time, 10 s orbital shaking at 200 rpm before measurement.

2.4. Preparation of Synthetic AI Standard

HHL and OHHL were purchased from Merck (Darmstadt, Germany). Stock solutions with a concentration of 20 mM were prepared in DMSO and stored at −20 °C. On the day of the bioassay, the stock was freshly diluted with sterile phosphate-buffered saline (PBS; 137 µM NaCl, 2.7 mM KCl, 8.1 mM Na_2HPO_4, 1.5 mM KH_2PO_4; pH 7.4) to concentrations ranging from 0.3 nM up to 200 nM and used as standard concentrations for calibration in the luminescence bioassay.

2.5. Cultivation of *Y. enterocolitica*

Y. enterocolitica strain DSM 11503 (biovar 3, serovar 0:9) was obtained from the DSMZ-German Collection of Microorganisms and Cell Cultures (Braunschweig, Germany). Overnight cultures were prepared by inoculating 5 mL tryptic soy broth (TSB, 17 g L^{-1} casein peptone, 3 g L^{-1} soy peptone, 2.5 g L^{-1} D(+) glucose, 5 g L^{-1} NaCl, 2.5 g L^{-1} K_2HPO_4; pH 7.3) with 5 µL of a thawed *Y. enterocolitica* glycerol stock. For experimental cultures, selective CIN medium was inoculated with the overnight culture at a ratio of 1:1000. The selective medium was derived from the CIN agar formulation as previously described [41]. Agar and color indicators were omitted from the preparation (17 g L^{-1} gelatin peptone, 1.5 g L^{-1} casein peptone, 1.5 g L^{-1} animal tissue peptone, 2 g L^{-1} yeast extract, 20 g L^{-1} mannitol, 0.5 g L^{-1} sodium deoxycholate, 0.5 g L^{-1} sodium cholate, 1 g L^{-1} sodium chloride, 2 g L^{-1} sodium pyruvate, 0.01 g L^{-1} magnesium sulfate, 0.004 g L^{-1} irgasan, 0.4 g L^{-1} cefsulodin, 0.25 g L^{-1} novobiocin; pH 7.4). All cultures were incubated at 28 °C, with shaking at 160 rpm.

2.5.1. Procedure of Growth Curve Analysis

A growth curve of *Y. enterocolitica* was recorded by inoculating 300 mL selective CIN medium with an overnight culture (1:1000) and taking 12 mL samples at regular intervals, starting at t = 0 and ending at 24 h. The optical density was measured at 600 nm (OD_{600}) to determine cell growth. Culture supernatants were generated from 10 mL aliquots (Section 2.5.2), and the cell number was estimated from the OD_{600} on the basis of a *Yersinia*-specific correlation factor determined in the context of this study by plating experiments. An OD_{600} value of 1.0 corresponds to $2.84 \cdot 10^8 \pm 3.9\%$ *Y. enterocolitica* cells mL^{-1}. The estimated cell concentration X was plotted semi-logarithmically against the cultivation time. The specific growth rate µ was determined by linear regression in the exponential phase. The specific growth rate µ of *Y. enterocolitica* and the doubling time t_D was calculated using the following equations [42]:

$$\mu = \frac{\ln X_t - \ln X_0}{t - t_0} \quad (1)$$

$$t_D = \frac{\ln 2}{\mu} \quad (2)$$

2.5.2. Preparation of Bacterial Culture Supernatants

Sterile culture supernatants were prepared from 10 mL samples taken during the growth curve experiment (Section 2.5.1). Bacterial cells were pelleted by centrifugation

(400× *g* for 5 min). The supernatants were filtered using a 0.2 μm syringe sterile filter (Carl Roth, Karlsruhe, Germany) and subsequently stored at −80 °C.

2.6. Extraction of AHLs

AHLs were extracted from sterile culture supernatants by liquid-liquid extraction (LLE) in pro-analysis-grade ethyl acetate (Carl Roth, Karlsruhe, Germany) containing 1% (*v*/*v*) analytical grade anhydrous acetic acid (Merck, Darmstadt, Germany). For this, acidified ethyl acetate was added to collected supernatant (1:3 *v*/*v*) and vortexed for 1 min. Afterwards, the organic phase was collected, and the remaining aqueous phase was extracted again as described. In total, five extraction cycles were performed. Pooled organic phases were concentrated by evaporation before drying using a SpeedVac vacuum concentrator (Eppendorf, Hamburg, Germany) for 1 h at RT. For LC-MS/MS analysis, the residue was reconstituted in 200 μL acetonitrile (LC-MS-grade, Carl Roth, Karlsruhe, Germany). To determine the performance and recovery rates achieved using the LLE protocol, we spiked the culture supernatants with 100 ng of OHHL and HHL as additional samples.

2.7. Analysis of Extracted AHLs Using LC-MS/MS

AHL extracts were passed through a Claristep 0.2 μm regenerated cellulose filter with an area of 0.7 cm^2 (Sartorius, Göttingen, Germany) into autosampler vials with inserts. The samples were separated by HPLC on a Curosil-PFP 3 μm 150 × 3.0 mm column (Phenomenex, Aschaffenburg, Germany) using the Agilent 1200 HPLC system with a diode array detector set to 190–400 nm. The mobile phases were (A) degassed ddH_2O containing 5 mM LiChropur ammonium acetate (Merck), 0.1% (*v/v*) HPLC-grade trifluoroacetic acid (Merck, Darmstadt, Germany), and 0.1% (*v/v*) ACS-grade formic acid (Carl Roth) and (B) ultra-LC-MS/MS-grade acetonitrile (Carl Roth, Karlsruhe, Germany) containing 0.1% (*v/v*) trifluoroacetic acid. We injected 10 μL samples and separated them by gradient flow at a constant flow rate of 0.4 mL min^{-1}. The gradient was held at 100% A for 2 min before changing to 100% B over a 12 min linear gradient, followed by a 2 min hold step at 100% B before returning to 100% A over 4 min, followed by a final re-equilibration for 5 min at 100% A. For calibration, stock solutions of OHHL and HHL (2 mg mL^{-1}, dissolved in DSMO) were thawed and diluted with acetonitrile to the following concentrations: 0.01, 0.05, 0.1, 0.15, 0.3, 0.5 μg mL^{-1}.

AHL extracts were passed through a Claristep 0.2 μm regenerated cellulose filter with an area of 0.7 cm^2 (Sartorius, Göttingen, Germany) into autosampler vials with inserts. The samples were separated by HPLC on a Curosil-PFP 3 μm 150 × 3.0 mm column (Phenomenex, Aschaffenburg, Germany) using the Agilent 1200 HPLC system with a diode array detector set to 190–400 nm. The mobile phases were (A) degassed ddH_2O containing 5 mM LiChropur ammonium acetate (Merck), 0.1% (*v/v*) HPLC-grade trifluoroacetic acid (Merck, Darmstadt, Germany), and 0.1% (*v/v*) ACS-grade formic acid (Carl Roth) and (B) ultra-LC-MS/MS-grade acetonitrile (Carl Roth, Karlsruhe, Germany) containing 0.1% (*v/v*) trifluoroacetic acid. We injected 10 μL samples and separated them by gradient flow at a constant flow rate of 0.4 mL min^{-1}. The gradient was held at 100% A for 2 min before changing to 100% B over a 12 min linear gradient, followed by a 2 min hold step at 100% B before returning to 100% A over 4 min, followed by a final re-equilibration for 5 min at 100% A. For calibration, stock solutions of OHHL and HHL (2 mg mL^{-1}, dissolved in DSMO) were thawed and diluted with acetonitrile to the following concentrations: 0.01, 0.05, 0.1, 0.15, 0.3, 0.5 μg mL^{-1}.

The HPLC fractions were analyzed on an API 3000/3200 Q TRAP device (Applied Biosystems, Darmstadt, Germany) equipped with an electro spray ionization probe. Mass scans were carried out in positive ion mode with the following parameters: curtain gas (30 AU), collision gas (medium), ion spray voltage (5500 V), source temperature (550 °C), ion source gas 1 (40), ion source gas 2 (50). AHLs were detected in two multiple reaction monitoring (MRM) experiments, one based upon the selective 102.1 Da lactone ring fragment shown previously to be well suited for AHL detection [43], and the other ion unique to the AHL under investigation. The optimized parameters for each AHL are described in the supplemental methods (Table S1).

2.8. Co-Cultivation of Y. enterocolitca and pMA Biosensor

Overnight cultures of the biosensor and *Y. enterocolitica* were prepared as described in Section 2.3 and 2.5. On the next day, 25 mL of CIN medium without cefsulodin, irgasan, or novobiocin was inoculated with the biosensor overnight culture at a ratio of 1:1000, and 20 mL of selective CIN medium was inoculated with the *Yersinia* overnight culture at the same ratio. The biosensor culture was incubated at 37 °C, shaking at 160 rpm, until an OD_{600} reached 0.4, at which point the culture was directly used in the bioassay. The *Yersinia* culture was incubated at 28 °C, shaking at 160 rpm, and was diluted to defined OD_{600} values once the OD_{600} had reached or slightly exceeded 0.05. These defined values were based on cell numbers previously selected for this study. The corresponding OD_{600} was estimated using the specific correlation factor (Section 2.5.1).

Similar to the method described in Section 2.3, 180 µL biosensor culture was mixed with 20 µL culture sample in each well of a C8 LockWell Lumi white 96-well plate (Thermo Fisher Scientific, Waltham, MA, USA). Kinetic measurement was carried out for 20 h at 30 °C, with detection of the luminescence signal every 20 min. The following parameters were set for the top optic measurement: 11.0 mm focal height, 10 s measuring interval, 10 s orbital shaking at 200 rpm before measurement, 0.1 s settling time. Optimized assays were conducted using the bottom optic measurement in a white 96-well microplate with a clear bottom and lid (Berthold Technologies, Bad Wildbad, Germany). Here, the focal height was set to −2.6 mm.

3. Results

3.1. Validation of pMA Biosensor

For evaluation of pMA biosensor functionality, a kinetic luminescence assay for five hours using standard concentrations of the synthetic OHHL AI in combination with pMA biosensor was performed (Figure 2a). PBS without OHHL was used as negative control (blank measurement). On the basis of the kinetic measurement, we were able to establish a correlation of the luminescence signals with different concentration of OHHL after 20, 40, 60, and 120 min by generating calibration curves (Figure 2b). The data were fitted by polynomial regression, and the minimal luminescence signal Lum_{min} was determined on the basis of Equation (3) [44]:

$$Lum_{min} = Average_{Blank\ value} + 3 \cdot SD_{Blank\ values} \quad (3)$$

By inserting Lum_{min} into corresponding regression function, the lowest detectable concentration of OHHL (LoD, limit of detection) could be calculated. Averaging the corresponding LoDs of three independent experiments, a LoD of 0.50 nM ± 0.29 nM OHHL could be calculated after 20 min of incubation. After further 20 min of incubation, the sensitivity had more than doubled, with a LoD of 0.21 nM ± 0.02 nM OHHL, while extension of the assay time to 60 min or 120 min produced even higher sensitivities of LoD = 0.20 nM ± 0.03 nM or 0.17 nM ± 0.05 nM, respectively.

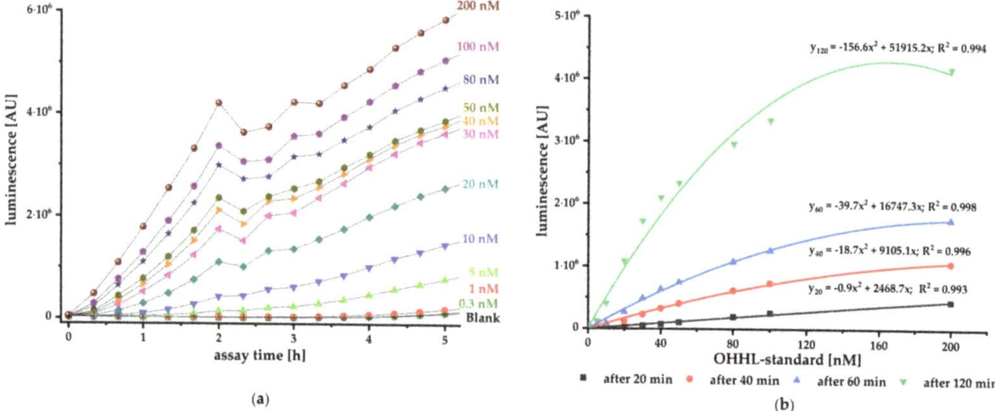

Figure 2. (**a**) Kinetic measurement of luminescence with the pMA biosensor and different standard concentrations of synthetic OHHL AI ranging from 200 nM to 0.3 nM (mean, $n = 3$). (**b**) Averaged calibration curves with corresponding regression functions at different times of luminescence measurement, indicating the relation between the signal generated by the pMA biosensor and OHHL standard concentrations (mean ± SD, $n = 3$).

3.2. Detection of AI from Bacterial Cultures

On the basis of our results using synthetic OHHL standards, we next focused on the detection of AIs produced by living bacteria. For this, *Y. enterocolitica*, producing OHHL and HHL, was cultivated for 24 h in selective CIN medium. Several samples throughout cultivation were collected for quantifying cell density and analyzing the presence of OHHL by using previously described kinetic luminescence assay. The growth curve of *Y. enterocolitica* is plotted in correlation with the luminescence signals obtained from the sterile filtered supernatant samples after 120 min of kinetic measurement (Figure 3). Although the increase in signal strength lagged slightly behind cell growth, a precise correlation between cell growth and increase in luminescence signal could be detected, demonstrating the potential of luminescence bioassay-based AI-detection. Additionally, a correlation of signal intensity and cell concentration could be detected, indicating the possibility of determining cell concentration with the presented bioassay.

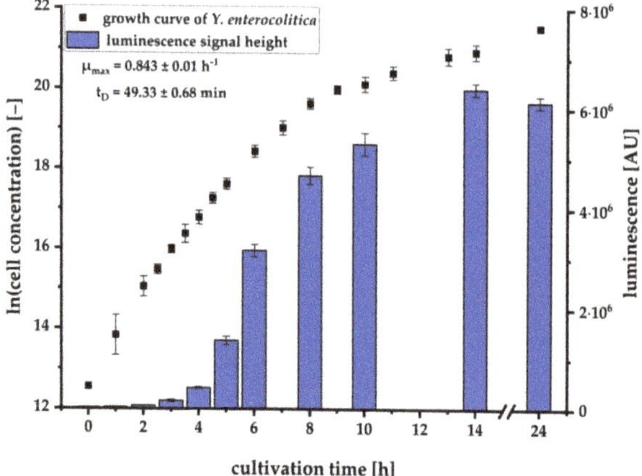

Figure 3. Luminescence signal height after 120 min (mean ± SD, $n = 3$) with the pMA biosensor and culture supernatants plotted against the corresponding growth curve (average of two independent experiments ± SD, $n = 4$, including specific growth rate μ and doubling time t_D).

As the luminescence signal could be induced by both types of AHLs, OHHL and HHL, the culture supernatants were analyzed regarding the specific AHL concentrations using LC-MS/MS. AHLs were obtained from the supernatants by liquid-liquid extraction and reconstitution in acetonitrile, achieving recoveries of 94.4 ± 9.6% for HHL and 99.2 ± 26.6% for OHHL for all processed samples (n = 10). AHL concentrations are compared to the corresponding growth curve in Figure 4a (gray and blue columns). As a result, a 3-6-fold excess of OHHL was detected compared to HHL, leading to the assumption that OHHL is mainly responsible for the luminescence signal observed in Figure 3. Therefore, the AI concentrations were calculated on the basis of OHHL calibration curves after 120 min and compared to the LC-MS/MS-based data in Figure 4a (green columns). A correlation between bacterial growth and the production of AI, which increased substantially during the exponential phase, could be revealed. Here, AI continued to accumulate until the stationary phase.

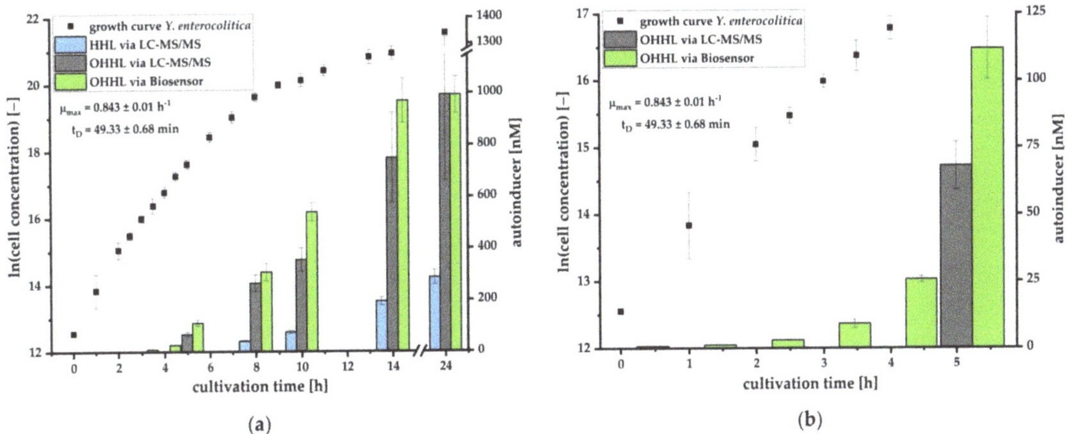

Figure 4. (**a**) Quantification of AI concentrations in culture supernatants of *Y. enterocolitica* by LC-MS/MS (OHHL: gray bars; HHL: blue bars; mean ± SD, n = 2) and by pMA biosensor (OHHL: green bars; mean ± SD, n = 3 calibration curves at 120 min assay time) in relation to the corresponding growth curve (average of two independent experiments ± SD, n = 4; including specific growth rate μ and doubling time t_D). (**b**) Zoomed section showing cultivation time 0 h to 6 h of cultivation.

An AI detection after a cultivation period of 5 h could be achieved using LC-MS/MS. However, the sensitivity was insufficient for detecting lower AI concentrations at earlier cultivation stages. In contrast, the biosensor-based assay reached sufficient sensitivity, and a detection of AHLs in the supernatant from early cultivation stages could be achieved (Figure 4b).

3.3. Co-Cultivation of *Y. enterocolitica* and pMA Biosensor

The sensitivity of the pMA biosensor-based assay was comparable to our established physico-chemical method based on LC-MS/MS. Although the sensitivity of the bioassay was considerably higher than our established LC-MS/MS protocol, small numbers of *Y. enterocolitica* cells cannot be detected without prior microbiological enrichment, and hence our rapid bioassay requires a short enrichment step. We therefore combined the *Y. enterocolitica* enrichment step with the kinetic measurement of the luminescence signal by co-cultivation. Culture samples containing different densities of proliferating *Y. enterocolitica* cells were added to the pMA biosensor, so both pathogen and biosensor were co-cultivated while the kinetic luminescence measurement was recorded.

In Figure 5, the initial signal increase is shown during the co-cultivation, demonstrating the cell number-dependent luminescence response of the pMA biosensor. High initial cell numbers resulted in an early increase of signal. Nevertheless, even the blank sample increased after 13.6 h of co-cultivation, which could be detected in several repetitions of the described experiment. However, the timing of blank-signal increase was not reproducible (data not shown). A distinction of blank samples from samples containing just a small number of cells might not be achievable, resulting in insufficient assay sensitivity. This observation could be explained by evaporation effects when assay plates were used that were not closed using a lid.

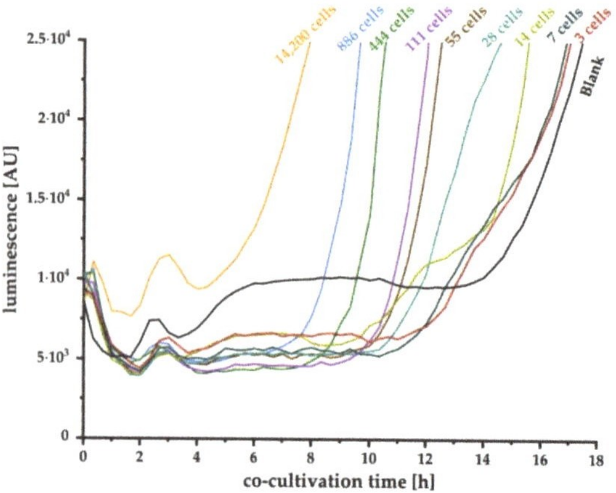

Figure 5. Kinetic measurement of luminescence recorded in intervals of 20 min during the co-cultivation of *Y. enterocolitica* and the pMA biosensor with readout direction from the top (detection plate without cover). Theoretical initial cell numbers (n = 3) were estimated using the OD_{600} and specific *Yersinia* correlation factor. The blank signal was obtained from cell medium without *Y. enterocolitica* (n = 11).

To counteract, the readout direction was changed to bottom optic, well gaps were filled with water, and the plate was closed using a fitting lid. The optimized method achieved a stable baseline for the blank sample, enabling reliable differentiation from low cell number samples (Figure 6a). A threshold was calculated using Equation (3) on the basis of the averaged blank signal, measured during the entire assay. Exceeding the threshold indicates a positive qualitative detection and the elapsed time before reaching the threshold provides a quantitative estimation of the initial cell number in the sample. As shown in Figure 6b, a correlation between cell number and time until passing the threshold could be detected. By this optimization, the sensitivity could be increased, resulting in a theoretical detection of a single cell *Y. enterocolitica* within a 20 µL sample after co-cultivation for 15.3 h.

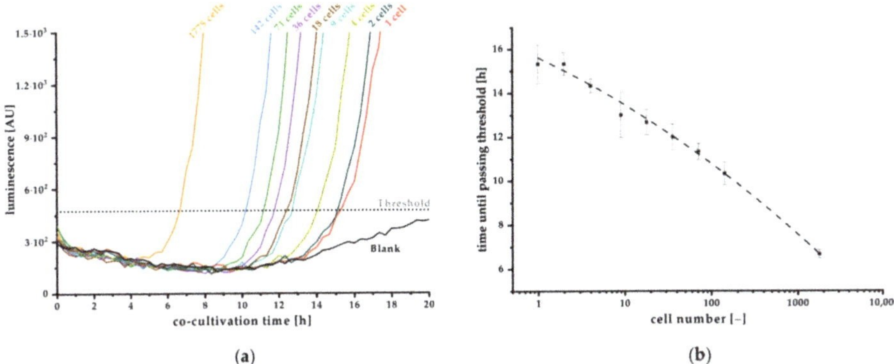

Figure 6. (**a**) Kinetic measurement of luminescence recorded in intervals of 20 min during the co-cultivation of *Y. enterocolitica* and the pMA biosensor according to the optimized co-cultivation approach with readout direction from the bottom (covered detection plate). A threshold was determined on the basis of the averaged blank in relation to the whole assay time and Equation (3). Exceeding the threshold indicates a positive, qualitative signal. Theoretical initial cell numbers ($n = 3$) were estimated using the OD_{600} and specific *Yersinia* correlation factor. The blank signal was obtained from cell medium without *Y. enterocolitica* ($n = 11$). (**b**) Correlation between initial cell number and time until threshold is exceeded fitted with Hill1 equation ($R^2 = 0.998$, $n = 3$), enabling a quantitative estimation of the initial cell number.

4. Discussion

We established and demonstrated a concept for rapid and sensitive detection of *Y. enterocolitica* by using QS. Our results proved the feasibility of a real-time detection by monitoring the production of the pathogen's AHLs using the pMA biosensor. Aiming to overcome the limitations of the conventional detection methods, we successfully adapted a co-cultivation approach to keep microbial enrichment and the simultaneous detection of *Y. enterocolitica* by the pMA biosensor as short and easy as possible.

To validate the pMA biosensor, we carried out kinetic luminescence assays using different standard concentrations of synthetic AI. The biosensor responded concentration-dependent to the authentic AHL and provides reliable and sensitive signals from 20 min onwards (Figure 2). It is particularly suitable for fast detection of small amounts OHHL with a sensitivity ranging from 0.50 nM up to 0.17 nM (LoD). This biosensor is at least as sensitive as the published constructs (pSB401, pSB403, pSB406, and pSB1075) with sensitivity to different AHLs in the pM and nM range [36].

We could further demonstrate that the pMA biosensor is able to respond to AHLs in culture supernatants of *Y. enterocolitica* (Figure 3) without being suppressed by compounds of selective media. The increase of the luminescence signal in relation to the growth of the pathogen demonstrates the characteristic accumulation of AHLs during cell growth. To investigate which AHL is produced during cell growth and is mainly responsible for induction of the *lux* cassette, we analyzed culture supernatants using LC-MS/MS. Here, we successfully verified the presence and production of OHHL and HHL during growth of *Y. enterocolitica*. OHHL is the cognate AI of LuxR and activates the *lux* gene expression most effectively [36]. HHL differs from OHHL by the absence of a carbonyl-group at the C3-atom in the side chain. This OHHL analogue binds with less efficiency to LuxR and induces the *lux* mechanism with less activity as OHHL. A previous study demonstrated that HHL inhibited the binding of OHHL to LuxR by about 10% when molar ratio was 1:1 of OHHL and HHL [45]. Our study revealed that *Y. enterocolitica* produced 3-6 fold less HHL than OHHL. Thus, an inhibition effect is negligible, confirming that the *lux* mechanism is primarily induced by OHHL. This condition allows for quantification of OHHL in culture supernatants by the pMA biosensor. Determined concentrations correlated in great manner with concentrations quantified by LC-MS/MS (Figure 4). The quantification by the pMA biosensor revealed a higher sensitivity than our LC-MS/MS-based analytical method, which by itself revealed a higher sensitivity than published methods demonstrating the ability for *Y. enterocolitica* detection in early stages of cultivation (compare [40]). Further research is necessary to determine whether and to what extent other AHLs may interact with the *lux* promoter. To generate data to the specificity or selectivity of the analytical procedure, we would need to investigate the selectivity of the selective medium and to determine which other bacteria could also produce OHHL and induce luminescent responses. This is especially important in terms of determining how specific our bacterial sensor is and what kind of selective enrichment steps need to be employed.

To date, several approaches have been developed using bacterial biosensors to detect and quantitatively estimate different kinds of bacteria on the basis of their secreted autoinducers [35]. However, sensitivity was not high enough to detect very low cell numbers. Hence, by applying the pMA biosensor in a co-cultivation approach, we could establish a real-time detection of *Y. enterocolitica* for the first time by monitoring produced OHHL. We further demonstrated that even single cell detection and reliable quantification is possible. Although the detection of small cell number requires several hours of co-cultivation, the presented method is considerably faster than classic microbiological cultivation techniques. The assay set up was successfully optimized, resulting in a stable blank signal throughout relevant duration of cultivation, facilitating the determination of a threshold value that defines the minimal luminescence signal, confirming a positive detection. With this approach, we achieved high sensitivity for detection of one initial cell in a sample after 15.3 h of co-cultivation. Hence, monitoring the luminescence signal and the time to exceed threshold also enables quantitative estimation of the initial cell concentration on the basis of the calibration curve plotted in Figure 6b. Here, the detection limit of one cell strongly depends on the reproducible low blank signal as realized in our improved experimental approach in combination with a reliable correlation factor between OD_{600} and cells per milliliter. The quantitative detection range is limited by high cell numbers, where autoinducer concentration is already high enough to induce expression of the *lux* operon. In this case, expression will start instantly, and strong luminescence signal increase can be observed within few minutes, as can be concluded from kinetics in Figure 2. Our established method is more time-effective than the conventional microbiological detection methods and ensures the detection of living cells as the release of AHLs is an active process. For a specific detection of *Y. enterocolitica*, further optimization has to be done, especially excluding luminescence signals induced by AIs from other bacteria. It is known that OHHL and similar AHLs that also might stimulate the *lux* expression cassette in the pMA biosensor are produced by a huge number of *Enterobacteria*. Therefore, a short pre-enrichment incubation step (e.g., 3 h) in CIN selection media might be beneficial to reduce the number of other bacteria and to mainly enrich *Y. enterocolitica*. However, CIN medium is not 100% selective to *Y. enterocolitica*, and therefore additional steps such as, e.g., an alkali treatment, could be necessary to increase selectivity [46,47]. Hence, further experiments have to be performed to reveal the specificity of our here presented assay when using real food samples. Nevertheless, our approach is well suited for fast an early detection of bacterial contamination and could be used in food security or veterinary applications. Further research will focus on the adaption of the luminescence cassette to specifically detect further AI molecules enabling fast detection of other pathogenic bacteria such as *Salmonella*, *Campylobacter*, or *Legionella*. Exchanging the luminescence gene by fluorescence cassettes expressing different fluorophores might additionally enable multiplex detection of different bacteria in a single sample.

Supplementary Materials: Supplementary materials can be found at https://www.mdpi.com/article/10.3390/bios11120517/s1. Figure S1. Plasmid map of sensor plasmid pMA-RQ_luxR_lux. It is based on the sequence of pSB401 [36], encoding a fusion construct combining the *Vibrio fischeri luxRI'* and *Photorhabdus luminescens luxCDABE* sequences. The fusion construct was synthesized and joined to the pMA-RQ backbone by GeneArt. The backbone carries an ampicillin resistance gene and a ColE1 origin of replication. Table S1. Optimized parameters for detection of N-hexanoyl- and N-(3-oxohexanoyl)-L-homoserine lactone using LC–MS/MS.

Author Contributions: Conceptualization: F.S. and M.F.; methodology: J.N., C.M. and A.C.; validation: J.N., A.C. and J.P.; formal analysis: J.N.; resources: F.S. and C.M.; investigation: J.N.; data curation: J.N.; writing—original draft preparation: J.N.; writing—review and editing: J.P., F.S., C.M. and A.C.; visualization: J.N.; supervision: F.S.; project administration: F.S. and M.F.; funding acquisition: F.S. and M.F. All authors have read and agreed to the published version of the manuscript.

Funding: This research was funded by the land North Rhine-Westphalia and the European Regional Development Fund (ERDF; German: EFRE) under grant number EFRE-0800966, reference number LS-1-2-005c.

Institutional Review Board Statement: Not applicable.

Informed Consent Statement: Not applicable.

Data Availability Statement: The datasets generated in this study are available from the corresponding author on request.

Acknowledgments: The authors would like to thank Richard Twyman for editorial assistance.

Conflicts of Interest: The authors declare no conflict of interest. The funders had no role in the design of the study; in the collection, analyses, or interpretation of data; in the writing of the manuscript; or in the decision to publish the results.

References

1. WHO; FAO. Joint FAO/WHO Expert Committee on Zooonoes: Second Report. In *World Health Organization Technical Report Series*; WHO: Geneva, Switzerland; FAO: Quebec City, QC, Canada, 1959.
2. Klous, G.; Huss, A.; Heederik, D.J.J.; Coutinho, R.A. Human-livestock contacts and their relationship to transmission of zoonotic pathogens, a systematic review of literature. *ONE Health* **2016**, *2*, 65–76. [CrossRef]
3. Abebe, E.; Gugsa, G.; Ahmed, M. Review on Major Food-Borne Zoonotic Bacterial Pathogens. *J. Trop. Med.* **2020**, *2020*, 4674235. [CrossRef] [PubMed]
4. European Food Safety Authority; European Centre for Disease Prevention and Control. The European Union One Health 2019 Zoonoses Report. *EFSA J.* **2021**, *19*, e06406. [CrossRef]
5. European Centre for Disease Prevention and Control. Yersiniosis: Annual Epidemiological Report for 2019. Available online: https://www.ecdc.europa.eu/sites/default/files/documents/AER-yersiniosis-2019_0.pdf (accessed on 30 October 2021).
6. Odyniec, M.; Stenzel, T.; Ławreszuk, D.; Bancerz-Kisiel, A. Bioserotypes, Virulence Markers, and Antimicrobial Susceptibility of Yersinia enterocolitica Strains Isolated from Free-Living Birds. *Biomed Res. Int.* **2020**, *2020*, 8936591. [CrossRef]
7. Bottone, E.J. Yersinia enterocolitica: Revisitation of an Enduring Human Pathogen. *Clin. Microbiol. Newsl.* **2015**, *37*, 1–8. [CrossRef]
8. Fàbrega, A.; Vila, J. Yersinia enterocolitica: Pathogenesis, virulence and antimicrobial resistance. *Enferm. Infecc. Microbiol. Clin.* **2012**, *30*, 24–32. [CrossRef] [PubMed]
9. Luciani, M.; Schirone, M.; Portanti, O.; Visciano, P.; Armillotta, G.; Tofalo, R.; Suzzi, G.; Sonsini, L.; Di Febo, T. Development of a rapid method for the detection of Yersinia enterocolitica serotype O:8 from food. *Food Microbiol.* **2018**, *73*, 85–92. [CrossRef]
10. Bottone, E.J. Yersinia enterocolitica: The charisma continues. *Clin. Microbiol. Rev.* **1997**, *10*, 257–276. [CrossRef]
11. Garzetti, D.; Susen, R.; Fruth, A.; Tietze, E.; Heesemann, J.; Rakin, A. A molecular scheme for Yersinia enterocolitica patho-serotyping derived from genome-wide analysis. *Int. J. Med. Microbiol.* **2014**, *304*, 275–283. [CrossRef]
12. Mazzette, R.; Fois, F.; Consolati, S.G.; Salza, S.; Tedde, T.; Soro, P.; Collu, C.; Ladu, D.; Virgilio, S.; Piras, F. Detection of Pathogenic Yersinia Enterocolitica in Slaughtered Pigs by Cultural Methods and Real-Time Polymerase Chain Reaction. *Ital. J. Food Saf.* **2015**, *4*, 4579. [CrossRef]
13. Van Damme, I.; Berkvens, D.; Vanantwerpen, G.; Baré, J.; Houf, K.; Wauters, G.; de Zutter, L. Contamination of freshly slaughtered pig carcasses with enteropathogenic Yersinia spp.: Distribution, quantification and identification of risk factors. *Int. J. Food Microbiol.* **2015**, *204*, 33–40. [CrossRef]
14. De Boer, E. Culture Media for the Isolation of Yersinia enterocolitica from Foods. In *Handbook of Culture Media for Food and Water Microbiology*, 3rd ed.; Corry, J.E.L., Baird, R.M., Curtis, G.D.W., Eds.; RSC Publishing: Cambridge, UK, 2012; Chapter 15; pp. 298–320, ISBN 978-1-84755-916-6.
15. EFSA. Monitoring and identification of human enteropathogenic *Yersinia* spp.—Scientific Opinion of the Panel on Biological Hazards. *EFSA J.* **2007**, *5*, 595. [CrossRef]
16. Petsios, S.; Fredriksson-Ahomaa, M.; Sakkas, H.; Papadopoulou, C. Conventional and molecular methods used in the detection and subtyping of Yersinia enterocolitica in food. *Int. J. Food Microbiol.* **2016**, *237*, 55–72. [CrossRef]
17. Thoerner, P.; Bin Kingombe, C.I.; Bögli-Stuber, K.; Bissig-Choisat, B.; Wassenaar, T.M.; Frey, J.; Jemmi, T. PCR detection of virulence genes in Yersinia enterocolitica and Yersinia pseudotuberculosis and investigation of virulence gene distribution. *Appl. Environ. Microbiol.* **2003**, *69*, 1810–1816. [CrossRef]

18. Vanantwerpen, G.; van Damme, I.; De Zutter, L.; Houf, K. Within-batch prevalence and quantification of human pathogenic Yersinia enterocolitica and Y. pseudotuberculosis in tonsils of pigs at slaughter. *Vet. Microbiol.* **2014**, *169*, 223–227. [CrossRef]
19. Jamali, H.; Paydar, M.; Radmehr, B.; Ismail, S. Prevalence, characterization, and antimicrobial resistance of Yersinia species and Yersinia enterocolitica isolated from raw milk in farm bulk tanks. *J. Dairy Sci.* **2015**, *98*, 798–803. [CrossRef]
20. Thibodeau, V. Development of an ELISA procedure to detect swine carriers of pathogenic Yersinia enterocolitica. *Vet. Microbiol.* **2001**, *82*, 249–259. [CrossRef]
21. Nielsen, B.; Heisel, C.; Wingstrand, A. Time course of the serological response to Yersinia enterocolitica O:3 in experimentally infected pigs. *Vet. Microbiol.* **1996**, *48*, 293–303. [CrossRef]
22. Ng, W.-L.; Bassler, B.L. Bacterial quorum-sensing network architectures. *Annu. Rev. Genet.* **2009**, *43*, 197–222. [CrossRef]
23. Rutherford, S.T.; Bassler, B.L. Bacterial quorum sensing: Its role in virulence and possibilities for its control. *Cold Spring Harb. Perspect. Med.* **2012**, *2*, a012427. [CrossRef]
24. Guo, M.; Gamby, S.; Zheng, Y.; Sintim, H.O. Small molecule inhibitors of AI-2 signaling in bacteria: State-of-the-art and future perspectives for anti-quorum sensing agents. *Int. J. Mol. Sci.* **2013**, *14*, 17694–17728. [CrossRef] [PubMed]
25. Federle, M.J.; Bassler, B.L. Interspecies communication in bacteria. *J. Clin. Investig.* **2003**, *112*, 1291–1299. [CrossRef] [PubMed]
26. Zhang, L.; Li, S.; Liu, X.; Wang, Z.; Jiang, M.; Wang, R.; Xie, L.; Liu, Q.; Xie, X.; Shang, D.; et al. Sensing of autoinducer-2 by functionally distinct receptors in prokaryotes. *Nat. Commun.* **2020**, *11*, 5371. [CrossRef] [PubMed]
27. Waters, C.M.; Bassler, B.L. QUORUM SENSING: Cell-to-Cell Communication in Bacteria. *Annu. Rev. Cell Dev. Biol.* **2005**, *21*, 319–346. [CrossRef] [PubMed]
28. Fuqua, C.; Parsek, M.R.; Greenberg, E.P. Regulation of gene expression by cell-to-cell communication: Acyl-homoserine lactone quorum sensing. *Annu. Rev. Genet.* **2001**, *35*, 439–468. [CrossRef]
29. Throup, J.P.; Camara, M.; Briggs, G.S.; Winson, M.K.; Chhabra, S.R.; Bycroft, B.W.; Williams, P.; Stewart, G.S. Characterisation of the yenI/yenR locus from Yersinia enterocolitica mediating the synthesis of two N-acylhomoserine lactone signal molecules. *Mol. Microbiol.* **1995**, *17*, 345–356. [CrossRef]
30. Whitehead, N.A.; Barnard, A.M.; Slater, H.; Simpson, N.J.; Salmond, G.P. Quorum-sensing in Gram-negative bacteria. *FEMS Microbiol. Rev.* **2001**, *25*, 365–404. [CrossRef]
31. Wang, S.; Payne, G.F.; Bentley, W.E. Quorum Sensing Communication: Molecularly Connecting Cells, Their Neighbors, and Even Devices. *Annu. Rev. Chem. Biomol. Eng.* **2020**, *11*, 447–468. [CrossRef]
32. Brodl, E.; Winkler, A.; Macheroux, P. Molecular Mechanisms of Bacterial Bioluminescence. *Comput. Struct. Biotechnol. J.* **2018**, *16*, 551–564. [CrossRef]
33. Steindler, L.; Venturi, V. Detection of quorum-sensing *N*-acyl homoserine lactone signal molecules by bacterial biosensors. *FEMS Microbiol. Lett.* **2007**, *266*, 1–9. [CrossRef]
34. Verbeke, F.; De Craemer, S.; Debunne, N.; Janssens, Y.; Wynendaele, E.; Van de Wiele, C.; De Spiegeleer, B. Peptides as Quorum Sensing Molecules: Measurement Techniques and Obtained Levels In vitro and In vivo. *Front. Neurosci.* **2017**, *11*, 183. [CrossRef]
35. Miller, C.; Gilmore, J. Detection of Quorum-Sensing Molecules for Pathogenic Molecules Using Cell-Based and Cell-Free Biosensors. *Antibiotics* **2020**, *9*, 259. [CrossRef]
36. Winson, M.K.; Swift, S.; Fish, L.; Throup, J.P.; Jørgensen, F.; Chhabra, S.R.; Bycroft, B.W.; Williams, P.; Stewart, G.S. Construction and analysis of luxCDABE-based plasmid sensors for investigating *N*-acyl homoserine lactone-mediated quorum sensing. *FEMS Microbiol. Lett.* **1998**, *163*, 185–192. [CrossRef]
37. Winson, M.K.; Swift, S.; Hill, P.J.; Sims, C.M.; Griesmayr, G.; Bycroft, B.W.; Williams, P.; Stewart, G.S. Engineering the luxCDABE genes from Photorhabdus luminescens to provide a bioluminescent reporter for constitutive and promoter probe plasmids and mini-Tn5 constructs. *FEMS Microbiol. Lett.* **1998**, *163*, 193–202. [CrossRef]
38. Bauer, J.S.; Hauck, N.; Christof, L.; Mehnaz, S.; Gust, B.; Gross, H. The Systematic Investigation of the Quorum Sensing System of the Biocontrol Strain Pseudomonas chlororaphis subsp. aurantiaca PB-St2 Unveils aurI to Be a Biosynthetic Origin for 3-Oxo-Homoserine Lactones. *PLoS ONE* **2016**, *11*, e0167002. [CrossRef]
39. Cataldi, T.R.I.; Bianco, G.; Palazzo, L.; Quaranta, V. Occurrence of N-acyl-L-homoserine lactones in extracts of some Gram-negative bacteria evaluated by gas chromatography-mass spectrometry. *Anal. Biochem.* **2007**, *361*, 226–235. [CrossRef]
40. Patel, N.M.; Moore, J.D.; Blackwell, H.E.; Amador-Noguez, D. Identification of Unanticipated and Novel N-Acyl L-Homoserine Lactones (AHLs) Using a Sensitive Non-Targeted LC-MS/MS Method. *PLoS ONE* **2016**, *11*, e0163469. [CrossRef]
41. Schiemann, D.A. Synthesis of a selective agar medium for Yersinia enterocolitica. *Can. J. Microbiol.* **1979**, *25*, 1298–1304. [CrossRef]
42. Fuchs, G.; Schlegel, H.G.; Eitinger, T. *Allgemeine Mikrobiologie*; Thieme: Stuttgart, Germany, 2007; ISBN 9783134446081.
43. Wang, J.; Ding, L.; Li, K.; Schmieder, W.; Geng, J.; Xu, K.; Zhang, Y.; Ren, H. Development of an extraction method and LC-MS analysis for N-acylated-l-homoserine lactones (AHLs) in wastewater treatment biofilms. *J. Chromatogr. B Analyt. Technol. Biomed. Life Sci.* **2017**, *1041–1042*, 37–44. [CrossRef]
44. Houba, V.J.G. *Guidelines for Quality Management in Soil and Plant Laboratories*; Food & Agriculture Organization: Rome, Italy, 1998; ISBN 9251040656.
45. Schaefer, A.L.; Hanzelka, B.L.; Eberhard, A.; Greenberg, E.P. Quorum sensing in Vibrio fischeri: Probing autoinducer-LuxR interactions with autoinducer analogs. *J. Bacteriol.* **1996**, *178*, 2897–2901. [CrossRef]

46. Tan, L.K.; Ooi, P.T.; Carniel, E.; Thong, K.L. Evaluation of a Modified Cefsulodin-Irgasan-Novobiocin Agar for Isolation of Yersinia spp. *PLoS ONE* **2014**, *9*, e106329. [CrossRef]
47. Aulisio, C.C.; Mehlman, I.J.; Sanders, A.C. Alkali method for rapid recovery of Yersinia enterocolitica and Yersinia pseudotuberculosis from foods. *Appl. Environ. Microbiol.* **1980**, *39*, 135–140. [CrossRef]

MDPI
St. Alban-Anlage 66
4052 Basel
Switzerland
www.mdpi.com

Biosensors Editorial Office
E-mail: biosensors@mdpi.com
www.mdpi.com/journal/biosensors

www.ingramcontent.com/pod-product-compliance
Lightning Source LLC
LaVergne TN
LVHW070654170726
843515LV00004B/409

* 9 7 8 3 0 3 6 5 9 0 8 2 0 *